Hands-on Deep Learning

Tanvir Islam

Hands-on Deep Learning

Building Models from Scratch

Tanvir Islam
Okta, Inc.
Bellevue, WA, USA

ISBN 978-3-032-00487-1 ISBN 978-3-032-00488-8 (eBook)
https://doi.org/10.1007/978-3-032-00488-8

This Springer imprint is published by the registered company Springer Nature Switzerland AG
The registered company address is: Gewerbestrasse 11, 6330 Cham, Switzerland

Preface

Building deep learning models can feel like navigating a complex maze. There are countless libraries, frameworks, and pre-built solutions that promise to simplify the process. But what if we want to truly understand the inner workings? What if we need to customize beyond the provided abstractions or troubleshoot a subtle bug that only surfaces at the foundational level?

This book is for those who crave that deeper understanding.

Hands-on Deep Learning: Building Models from Scratch isn't merely about calling an API. It's about getting hands-on with the fundamental mathematical and algorithmic principles that underpin deep learning. We'll strip away the layers of abstraction and guide ourselves through the process of constructing neural networks, layer by layer, using only the most basic building blocks. We'll implement activation functions, optimize weights with gradient descent, and even build convolutional and recurrent layers from the ground up.

By tackling deep learning in this hands-on, from-scratch manner, we'll gain an intuitive grasp of how these powerful models learn, make predictions, and sometimes, struggle. This knowledge will not only make us more proficient deep learning practitioners, but also more confident and adaptable ones, ready to tackle novel problems and push the boundaries of what's possible together.

This book is designed for data scientists and machine learning engineers who are keen to dive deep into the complexities of deep learning. It's a comprehensive guide for anyone eager to explore and deepen their knowledge in this field. Furthermore, undergraduate and graduate students will find this book to be a valuable resource.

We'll begin with the essentials—matrix multiplication, loss metrics, and gradient-descent optimization—and progressively build toward complex architectures: from multilayer perceptrons to convolutional networks and recurrent models. At each stage, we'll pause to write concise, self-contained modules, codes, and functions; then integrate them into complete training pipelines that solidify both theory and practice.

In a nutshell, this book is an invitation to think like a deep learning engineer, not just a user of deep learning tools. By building every component from scratch—and applying each concept to real-world examples—we gain an authentic, working understanding of how deep learning algorithms work in practice. Through hands-on projects and practical case studies, we'll see how theoretical principles translate into solutions for real problems. The goal isn't simply to train networks—it's to *understand*. This is where deep learning stops being *magic* and starts being *mastery*.

Prerequisites

- Python (intermediate)
  - Has worked with Python syntax and data structures.
  - Has used Python libraries such as NumPy and Pandas in projects or tasks.
  - Has experience in Python programming.
- Machine Learning knowledge (basic)
  - Has an understanding of fundamental machine learning concepts such as supervised and unsupervised learning.
  - Has familiarity with basic machine learning workflow, such as training, validation, and data wrangling.
  - Has experience with a machine learning framework such as scikit-learn and TensorFlow.
- Linear algebra and calculus (intermediate)
  - Has an understanding of intermediate linear algebra concepts such as vectors, matrices, and operations on them.
  - Has familiarity with intermediate calculus concepts such as differentiation and integration.
  - Has the ability to apply these mathematical concepts in the context of machine learning and deep learning.

Takeaways

- Gain a solid understanding of how to build a deep learning model from scratch.
- Learn how to implement the Gradient Descent algorithm, both in its non-vectorized and vectorized forms, and understand its role in training deep neural networks.
- Master the process of creating a neural network, including implementing neural network representations, activation functions, forward and backward propagation, and cost function.
- Deal with bias and variance in machine learning models and learn various regularization techniques, including dropout and other methods.

- Leverage advanced optimization techniques such as Gradient Descent with Momentum, RMSProp, and Adam for efficient training of deep learning models.
- Apply Convolutional Neural Networks (CNNs) in computer vision tasks through convolutions, padding, pooling, and stride.
- Implement Recurrent Neural Networks (RNNs) and Long Short-Term Memory Networks (LSTMs), powerful tools for processing sequential data, with practical examples like creative text generation and text auto-completion.
- Use embeddings in sequence models for tasks like topic classification, understanding key concepts like the embedding matrix, cosine similarity, beam search, and the Bleu score.
- Implement the Attention Mechanism and Transformers, advanced techniques that have revolutionized the field of natural language processing.

Acknowledgments

It is with immense gratitude that I acknowledge the foundational contributions that have shaped this book. Andrew Ng's accessible machine learning and deep learning courses provided the initial spark and laid a robust pedagogical framework. The pioneering work of Geoffrey Hinton, Yoshua Bengio, and Yann LeCun, as encapsulated in their groundbreaking papers, has been an endless source of inspiration. Finally, François Chollet's book and Google's tutorials served as an invaluable practical guide, illuminating the path to hands-on implementation with TensorFlow. This book stands on the shoulders of their collective contributions, and I hope it will help others embark on their own hands-on journey into deep learning.

Bellevue, WA, USA Tanvir Islam

Competing Interests The author has no competing interests to declare that are relevant to the content of this manuscript.

Contents

Chapter 1
Introduction

This chapter covers:

- Defining artificial intelligence (AI) and deep learning.
- Exploring the hierarchical relationship between AI and deep learning.
- Outlining the mental model of building deep learning models.
- Identifying the key elements required for effective model development.
- Explaining the core building blocks of deep learning models.

In recent years, deep learning has emerged as a groundbreaking technology (LeCun et al. 2015), revolutionizing various industries (Krizhevsky et al. 2012) and reshaping the way we interact with the world. Deep learning has demonstrated its potential to solve complex problems and drive innovation in various ways. In the medical field, it assists in diagnosing diseases (Esteva et al. 2017), forecasting patient outcomes, and tailoring treatments by examining medical images and identifying irregularities. Self-driving cars use deep learning for object detection (Bojarski et al. 2016), route planning, and decision-making, ensuring safe travel. Natural language processing tools, such as chatbots and virtual assistants, employ deep learning to answer questions in human language (Brown et al. 2020). In the financial sector, it aids in detecting fraud, managing risks, and executing algorithmic trades by processing vast amounts of data (Heaton et al. 2016). Streaming platforms such as Netflix and Spotify use deep learning to offer personalized recommendations (Gomez-Uribe and Hunt 2015). E-commerce platforms (Linden et al. 2003) also provide personalized recommendations based on user preferences and browsing history. Furthermore, deep learning boosts security through facial recognition and cybersecurity measures, and aids agriculture by monitoring crops, forecasting yields, and identifying crop diseases.

Now, the question is: how to build a deep learning model? There is no universal method for building a deep learning model. Each model needs to be tailored to the specific application or use case. The development process is typically experimental and iterative, necessitating a thorough understanding of both the problem domain and the foundational algorithms. As a result, building an effective deep learning

T. Islam, *Hands-on Deep Learning*, https://doi.org/10.1007/978-3-032-00488-8_1

model is both an art and a science, requiring creativity, expertise, and ongoing refinement.

Beginning with this chapter, we set the foundation for understanding the mental models of building deep learning models. The chapter starts with clear definitions of artificial intelligence (AI) and deep learning (DL). It then looks into the hierarchical relationship between AI, machine learning (ML), and deep learning, highlighting how these fields interconnect. It provides the big picture of building deep learning models, emphasizing the interconnected elements. Additionally, the chapter also covers the essential building blocks of a deep learning model. This chapter will set the stage to build deep learning models from scratch.

1.1 Artificial Intelligence and Deep Learning

The term "artificial intelligence" (AI) (Russell and Norvig 2016) has gained significant attention in recent years. But what is artificial intelligence or AI? To be fair, "artificial intelligence" is not a novel term. It is being used in various sectors for decades to solve a variety of problems. The concept of AI is broad and diverse, with a plethora of definitions found in scholarly literature. In this context, we will go through two prominent definitions.

The first one, as defined by Oxford Languages, is:

> Artificial Intelligence refers to the theory and development of computer systems able to perform tasks that normally require human intelligence, such as visual perception, speech recognition, decision-making, and translation between languages.

On the other hand, Google provides a slightly different perspective:

> Artificial intelligence is a field of science concerned with building computers and machines that can reason, learn, and act in such a way that would normally require human intelligence or that involves data whose scale exceeds what humans can analyze.

These definitions, while distinct, converge on the idea that AI is about replicating or even surpassing aspects of human intelligence using machines. It is a domain that keeps growing and improving, constantly testing what machines can do.

Artificial intelligence (AI) is a broad field with various subfields (Fig. 1.1). One significant subfield is machine learning (ML), which focuses on creating computer systems that can learn from data and enhance their abilities without needing specific programming. Within ML, deep Learning is a more specialized field that uses multilayered neural networks (which is why it is called "deep") to model and understand complex patterns in datasets (Goodfellow 2016). It is like teaching a computer to learn through layers of information.

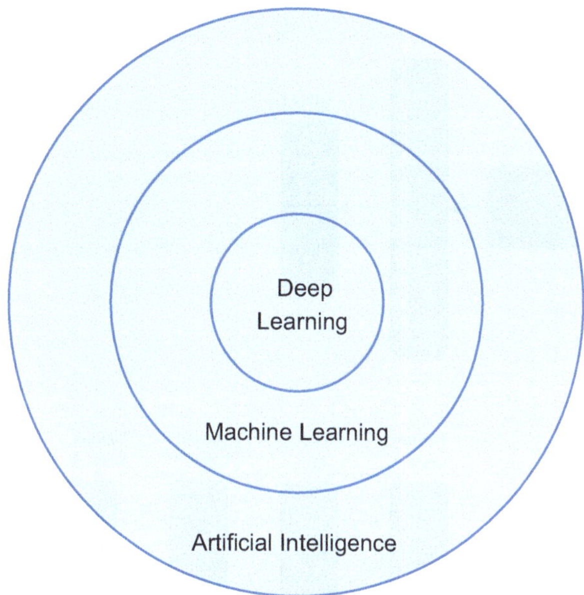

Fig. 1.1 The hierarchical relationship of deep Learning within artificial intelligence (AI)

1.1.1 Neural Networks

Deep learning uses neural networks (Goodfellow 2016). A neural network is a computational architecture that draws inspiration from the operational mechanisms of the human brain (McCulloch and Pitts 1943). A neural network is comprised of many layers, containing an input layer, one or more hidden layers, and an output layer. A high-level illustration of neural networks is shown in Fig. 1.2. The layers contain nodes or artificial neurons, and they are interconnected. Every link between nodes carries a weight, and the network adapts by modifying these weights in response to input data and the expected output. Neural networks are designed to recognize patterns, learn from data, and make intelligent decisions or predictions (Rosenblatt 1958). We may not yet have a full understanding of all these terminologies, but we will discuss them in detail over the next two chapters.

1.1.2 Deep Neural Networks

A deep neural network refers to a neural network that possesses a substantial number of layers nested between the input and output layers (Montufar et al. 2014), as illustrated in Fig. 1.2 (bottom panel). The deep neural networks term refers to stacking of the layers in a sequential manner, where each layer feeds its output into the next layer as input. The term "depth" in this context defines whether the neural network is classified as a deep neural network or a shallow neural network.

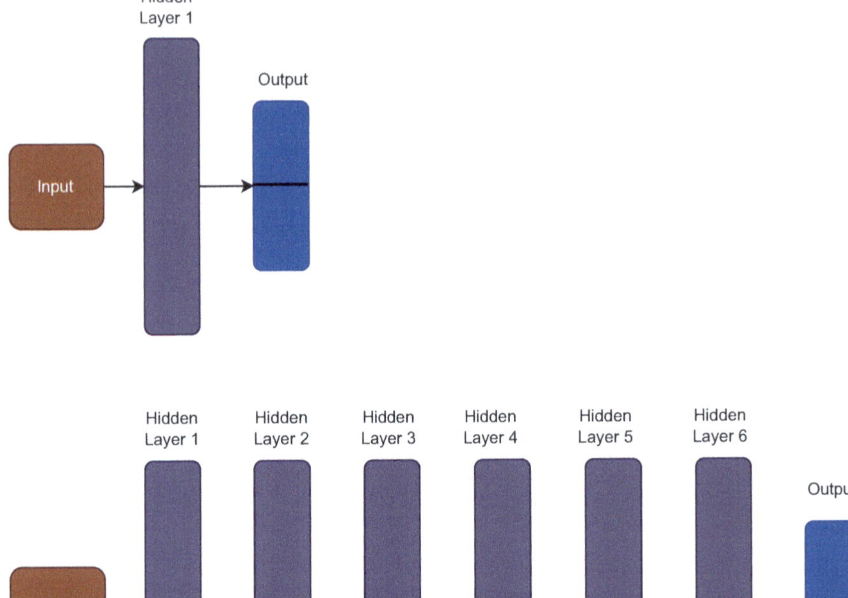

Fig. 1.2 This figure provides a visual comparison between a traditional neural network and a deep neural network. The traditional neural network consists of an input layer, one hidden layer, and an output layer. In contrast, the deep neural network features multiple hidden layers between the input and output layers

Generally speaking, a neural network with a depth of two or more hidden layers can be considered as a deep neural network.

Size is important to deep neural networks in two key ways (see Fig. 1.3):

- The performance of a neural network tends to improve as its size increases (Kaplan et al. 2020), especially when trained with a large dataset. This is because larger neural networks have more parameters (weights and biases). This means it has more capacity to learn complex patterns, especially with a large dataset.
- If the training set is small, nonneural network algorithms might outperform neural networks, particularly when careful feature engineering is applied. This is because with a small dataset, neural networks can easily memorize the training data, leading to overfitting (Srivastava et al. 2014). This means they perform well on the training data but poorly on unseen data.

We have also seen the surge of the use of deep learning models in modern-day applications. This surge in deep learning can be traced back to several key factors:

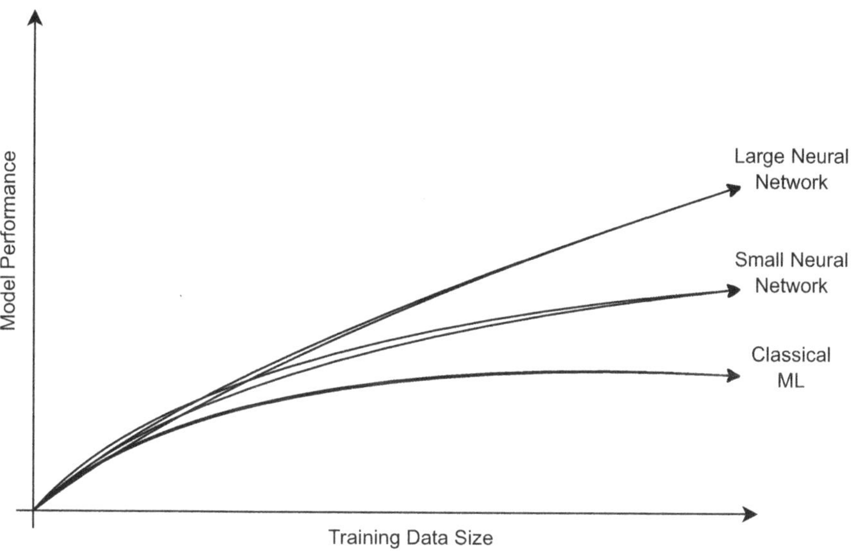

Fig. 1.3 The influence of neural network size on model performance in modern-day applications. Larger networks, with more layers and neurons, have the potential to capture complex patterns and relationships within data, leading to improved accuracy and performance. However, this requires more training data

- The availability of substantial amounts of data (Hilbert and López 2011).
- The advent of cloud computing, which provides immense computational power (Dean et al. 2012; Jouppi et al. 2017).
- The considerable contributions to research and the competitive landscape among leading tech companies, often referred to as FAANG (Facebook, Amazon, Apple, Netflix, and Google).

1.2 Big Picture

Building a deep learning model is not just about the model itself. We also have to think other three important elements shown in Fig. 1.4 while building deep learning models. These elements are the problem we are attempting to solve, the data we have, and the computational resources we require. All these things are connected and we need to consider them all together.

Let us look into these:

1. **Problem Statement:**

The first step in building a deep learning model is to clearly define the problem we want to solve. This involves setting specific, measurable goals and determining if deep learning is the right approach for the problem. Not all problems are suitable for

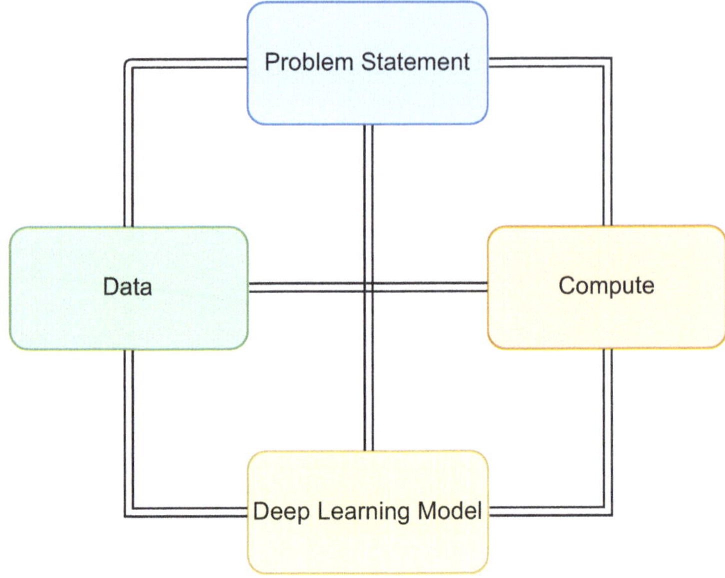

Fig. 1.4 Big picture of building deep learning models

deep learning, so it is important to assess this early on (Domingos 2012). For example, in some cases, domain-specific algorithms, heuristics, or classic machine learning might be more effective than deep learning. If deep learning is the right approach for the problem, the next step is to formulate the problem statement. Should we frame it as a tabular classification problem, an image classification problem, a regression problem, or something else? This problem statement will help us to determine what type of deep learning model we will have to build, given the data at hand, which is another element.

2. **Data**:

Data is the fuel for any deep learning model. It is the information that the model will learn from. The performance of the model is directly impacted by the quality of the data. Therefore, gathering relevant and diverse data is crucial. Data will also steer us to the right direction and help us determine the type of deep learning model we need to build.

Deep learning can be applied to both structured and unstructured data:

- *Structured data*: Structured data includes tabular data. Examples include house price prediction, ad-click prediction, and so on.
- *Unstructured data*: Unstructured data includes audio, images, and text. Examples include image recognition, speech recognition, natural language understanding, and so on.

3. **Compute Resources:**

Deep learning models, especially those with large amounts of data, require significant computational power and storage. This includes hardware such as GPUs for training the models and sufficient memory to store the data and the models. Availability of compute resources will also help us determining what type of deep learning model we will have to build. For instance, if our resources are limited, we might consider exploring alternative architectures that are less resource-intensive yet still effective. The key is to strike a balance between model complexity and computational feasibility with the cost in mind.

4. **Deep Learning Model:**

Our final element is obviously the deep learning model itself. However, as discussed, all of the above three elements play crucial role in building a deep learning model. The consideration of all these elements will define what type of model we will have to build. Deep learning model is essentially the architecture or the algorithm that learns from the data to solve the problem defined in the problem statement. It could be a convolutional neural network for image recognition tasks, a recurrent neural network for sequence data, or any other suitable model. Compute resources facilitate the model building or training process.

In essence, building an effective deep learning model is a holistic process based on the other three interlinked elements. Each element is integral to the model's success and cannot be overlooked. Understanding this big picture is fundamental to embarking on the journey of building an effective deep learning model.

1.3 Deep Learning Model Building Blocks

Now we know the big picture. Given a problem statement, data at hand, and compute resources at our disposal, we will be determining what type of model need to be built, and how. There are various ways of building an efficient deep learning model. There are many parts. There are many choices to make. Figure 1.5 provides the core building blocks to build a deep learning model. For the sake of understanding, they are divided into four core blocks. The first block is optimization algorithms, the

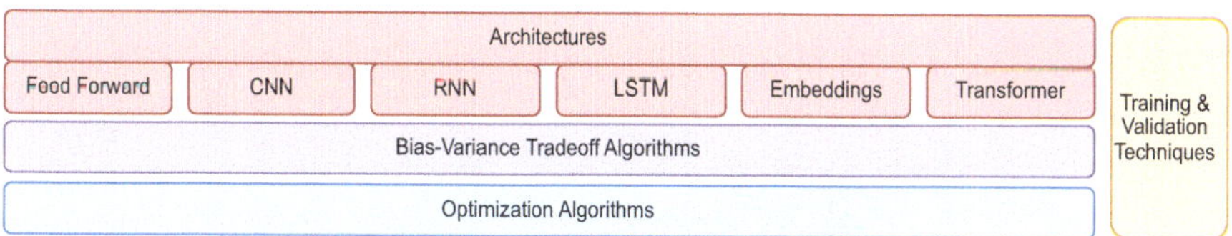

Fig. 1.5 The core building blocks to build a deep learning model. This figure outlines the fundamental components necessary for building a deep learning model, divided into four core blocks for clarity. Each of these blocks plays a crucial role in the overall process of building an effective deep learning model

second block is bias–variance trade-off algorithms, the third block is neural network architectures, and the fourth block is training and validation techniques.

Building an effective deep learning model is analogous to preparing a good recipe. Just as in cooking, building a deep learning model requires a variety of essential elements. The optimization algorithms, such as gradient descent (Rumelhart et al. 1986), might be analogous to the burner. They provide the heat source that drives the cooking process, or in this case, the learning process. The settings of the burner, such as the learning rate in optimization, determine how fast or slow the cooking happens. The architectures of the models are akin to the ingredients in a recipe. For instance, in a computer vision problem, we might use convolutional neural networks (CNNs) (He et al. 2016) just as we might use beef and cheese in a cheesesteak. The choice and combination of ingredients can drastically change the flavor of the dish, just as the choice of architecture can significantly affect the performance of the model. Different deep learning architectures or their combinations are suitable for different types of problems, and not every architecture will work well for every kind of problem. The bias–variance trade-off algorithms could be compared to the seasoning in cooking. Just as too much or too little seasoning can ruin a dish, overfitting (high variance) or underfitting (high bias) can lead to poor model performance. Striking the right balance is key in both scenarios. Finally, the act of training the model is like cooking the entire dish. It is the process where all the ingredients come together under the heat of the burner, transforming into something new and delicious. Similarly, the model is trained using an optimization algorithm that leverages a specific architecture or a combination of architectures, through training and validation techniques, to ensure the bias–variance trade-off is satisfied.

These building blocks are essential for building a deep learning model, as illustrated in Fig. 1.6. Problem statements and data will guide us in defining a

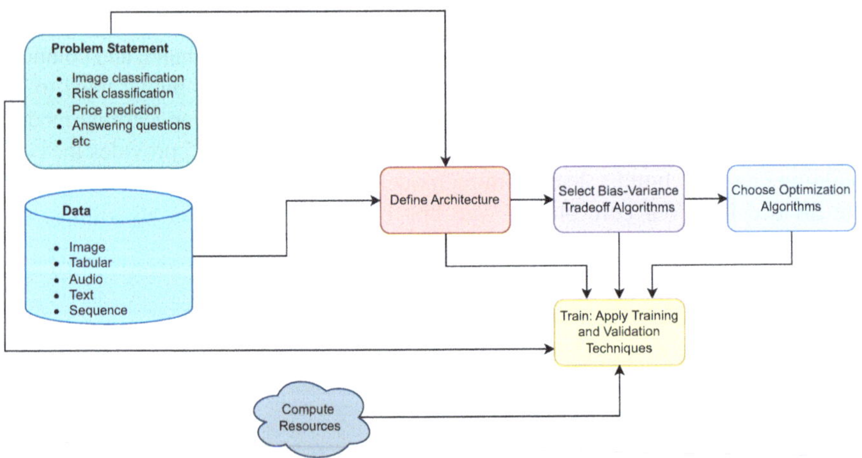

Fig. 1.6 High-level overview of building a deep learning model. This figure illustrates the key stages involved in building a deep learning model, from problem formulation and data collection to model training and validation

suitable deep learning architecture. We will then select appropriate algorithms to manage the bias–variance trade-off (Geman et al. 1992) and integrate them into the architecture. Additionally, we will choose effective optimization algorithm (Kingma 2014), for example, a variant of gradient descent. Finally, we will train the model using various training and validation techniques.

In summary, building a deep learning model is a lot like cooking a recipe. It involves picking and mixing the right ingredients (architectures), adjusting the stove's heat (optimization algorithms), adding the perfect amount of spices (bias–variance trade-off), and using the correct cooking method (training) to make it all work. Within each block, there are many choices to make. The art of deep learning is in choosing the right components for the task at hand and tuning them to work together in the best possible way. A good understanding of *all* these blocks and choices will help us building effective deep learning models. In this book, we will be covering all these building blocks and choices. Knowing all these building blocks and choices will help us make good decisions about where to use which blocks, and how.

This book will follow a structured format. First, we will introduce the concept and underlying theory. As we discuss the theory, we will implement the corresponding codes and functions. We will write these codes in the context of solving a real-world problem. We will follow this format for all the building blocks. The book will start from first principles using NumPy and gradually transition to TensorFlow in later chapters.

Optimization algorithms is the foundational building block. Specifically, gradient descent is the fundamental algorithm (Rumelhart et al. 1986) used for optimization, in order to build deep learning models. In the next chapter, we will cover this essential building block in detail.

1.4 Summary

- AI encompasses multiple subfields, including machine learning (ML), which focuses on systems that learn from data without explicit programming.
- Deep learning, a subfield of machine learning, uses multilayered neural networks to model and understand complex patterns in data.
- Neural networks, which draw inspiration from the human brain, consist of interconnected layers of nodes that adapt based on input data and expected output.
- Building a deep learning model involves formulating a clear problem statement, gathering quality data, and ensuring adequate computational resources.
- Effective deep learning models require a balance of optimization algorithms, neural network architectures, bias–variance trade-offs, and appropriate training and validation techniques.

References

Bojarski, M., Del Testa, D., Dworakowski, D., Firner, B., Flepp, B., Goyal, P., Jackel, L. D., Monfort, M., Muller, U., Zhang, J., & others. (2016). End to end learning for self-driving cars. *arXiv Preprint arXiv:1604.07316.*

Brown, T., Mann, B., Ryder, N., Subbiah, M., Kaplan, J. D., Dhariwal, P., Neelakantan, A., Shyam, P., Sastry, G., Askell, A., & others. (2020). Language models are few-shot learners. *Advances in Neural Information Processing Systems, 33*, 1877–1901.

Dean, J., Corrado, G., Monga, R., Chen, K., Devin, M., Mao, M., Ranzato, M., Senior, A., Tucker, P., Yang, K., & others. (2012). Large scale distributed deep networks. *Advances in Neural Information Processing Systems, 25.*

Domingos, P. (2012). A few useful things to know about machine learning. *Communications of the ACM, 55*(10), 78–87.

Esteva, A., Kuprel, B., Novoa, R. A., Ko, J., Swetter, S. M., Blau, H. M., & Thrun, S. (2017). Dermatologist-level classification of skin cancer with deep neural networks. *Nature, 542*(7639), 115–118.

Geman, S., Bienenstock, E., & Doursat, R. (1992). Neural networks and the bias/variance dilemma. *Neural Computation, 4*(1), 1–58.

Gomez-Uribe, C. A., & Hunt, N. (2015). The Netflix recommender system: Algorithms, business value, and innovation. *ACM Transactions on Management Information Systems (TMIS), 6*(4), 1–19.

Goodfellow, I. (2016). *Deep learning.* MIT press.

He, K., Zhang, X., Ren, S., & Sun, J. (2016). Deep residual learning for image recognition. *Proceedings of the IEEE Conference on Computer Vision and Pattern Recognition*, 770–778.

Heaton, J., Polson, N. G., & Witte, J. H. (2016). Deep learning in finance. *arXiv Preprint arXiv:1602.06561.*

Hilbert, M., & López, P. (2011). The world's technological capacity to store, communicate, and compute information. *Science, 332*(6025), 60–65.

Jouppi, N. P., Young, C., Patil, N., Patterson, D., Agrawal, G., Bajwa, R., Bates, S., Bhatia, S., Boden, N., Borchers, A., & others. (2017). In-datacenter performance analysis of a tensor processing unit. *Proceedings of the 44th Annual International Symposium on Computer Architecture*, 1–12.

Kaplan, J., McCandlish, S., Henighan, T., Brown, T. B., Chess, B., Child, R., Gray, S., Radford, A., Wu, J., & Amodei, D. (2020). Scaling laws for neural language models. *arXiv Preprint arXiv:2001.08361.*

Kingma, D. P. (2014). Adam: A method for stochastic optimization. *arXiv Preprint arXiv:1412.6980.*

Krizhevsky, A., Sutskever, I., & Hinton, G. E. (2012). ImageNet classification with deep convolutional neural networks. *Advances in Neural Information Processing Systems, 25.*

LeCun, Y., Bengio, Y., & Hinton, G. (2015). Deep learning. *Nature, 521*(7553), 436–444.

Linden, G., Smith, B., & York, J. (2003). Amazon. Com recommendations: Item-to-item collaborative filtering. *IEEE Internet Computing, 7*(1), 76–80.

McCulloch, W. S., & Pitts, W. (1943). A logical calculus of the ideas immanent in nervous activity. *The Bulletin of Mathematical Biophysics, 5*, 115–133.

Montufar, G. F., Pascanu, R., Cho, K., & Bengio, Y. (2014). On the number of linear regions of deep neural networks. *Advances in Neural Information Processing Systems, 27.*

Rosenblatt, F. (1958). The perceptron: A probabilistic model for information storage and organization in the brain. *Psychological Review, 65*(6), 386.

Rumelhart, D. E., Hinton, G. E., & Williams, R. J. (1986). Learning representations by back-propagating errors. *Nature*, *323*(6088), 533–536.

Russell, S. J., & Norvig, P. (2016). *Artificial intelligence: A modern approach*. Pearson.

Srivastava, N., Hinton, G., Krizhevsky, A., Sutskever, I., & Salakhutdinov, R. (2014). Dropout: A simple way to prevent neural networks from overfitting. *The Journal of Machine Learning Research*, *15*(1), 1929–1958.

Chapter 2
Implementing the Gradient Descent Algorithm

This chapter covers:

- Understanding gradient descent in optimizing parameters for neural networks.
- Exploring logistic regression to grasp gradient descent concepts.
- Implementing gradient descent step-by-step.
- Visiting both non-vectorized and vectorized approaches of gradient descent.

In the previous chapter, we covered the core building blocks of deep neural networks. One of these essential building blocks in training neural networks is optimization algorithms. Gradient descent is the foundational building block for training neural networks (Rumelhart et al., 1986). Specifically, the key to training these networks lies in optimizing their parameters (Nocedal & Wright, 1999), and the gradient descent algorithm plays a crucial role in this process. This chapter explains how gradient descent works, focusing on its importance in minimizing loss functions (Bishop & Nasrabadi, 2006) to obtain the parameters of neural networks. We will start with the basics of logistic regression to build a foundation, then move on to the step-by-step implementation of gradient descent, both in a simple, non-vectorized form and a more efficient, vectorized form. This chapter serves as the starting point, with gradient descent being the key component in building deep neural network models from scratch.

2.1 Introduction to Gradient Descent

A neural network fundamentally operates on the principle of function mapping (Goodfellow, 2016). In essence, it takes a set of feature values and, once trained, maps this feature space, denoted as X, into a corresponding target space, y. For example, consider a scenario where the feature values include the size of a house and the number of bedrooms it contains. A well-trained neural network should be capable of mapping these features into a target space that accurately predicts the

T. Islam, *Hands-on Deep Learning*, https://doi.org/10.1007/978-3-032-00488-8_2

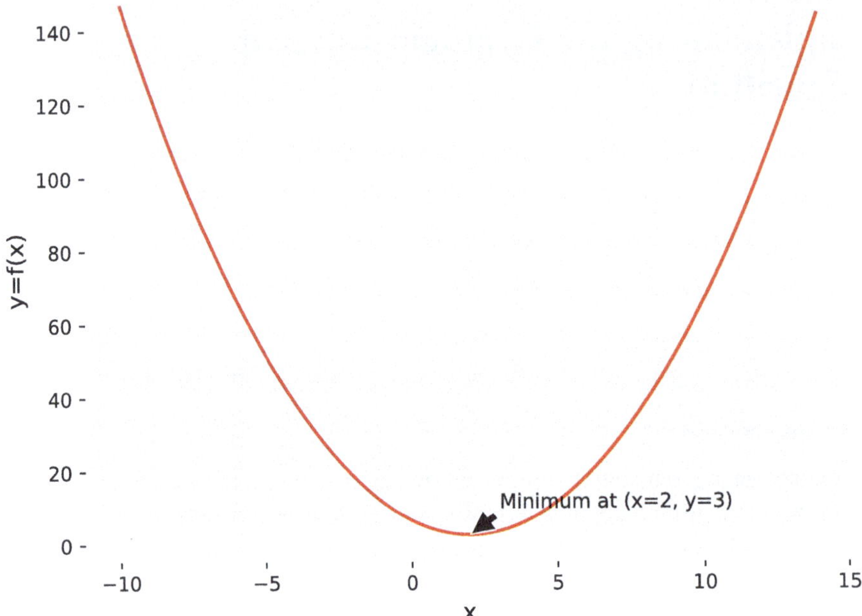

Fig. 2.1 Finding minimum of an objective function through gradient descent

house's price. This process is commonly referred to as prediction. However, we may wonder: how does the network learn this mapping? For the sake of discussion, let us assume we have access to historical observations. But how can we develop a mapping function based on these historical observations? This is where gradient descent algorithm comes into play, a fundamental building block of deep neural networks.

Before diving into the loss function or function mapping, let us develop some intuition first on gradient descent. Given a differentiable function, the primary objective of the gradient descent algorithm is to find the minimum value of the function (Boyd & Vandenberghe, 2004). Consider an arbitrary function $y = (x - 2)^2 + 3$ as an example. Our goal is to determine the value of x at which this function reaches its minimum. It is important to note that we are interested in the function's minimum, which does not necessarily have to be zero. Generally, a function's minimum point is where its derivative is zero. For instance, the minimum of this function is at $x = 2$ (Fig. 2.1).

The gradient descent algorithm is designed to find the minima of any differentiable function. It seeks to answer the question: at what value of the parameters does the function reach its minimum? In this example, our arbitrary function is $y = (x - 2)^2 + 3$, and we are interested in finding the value of x that minimizes y.

When it comes to the training of neural networks, this function is referred to as a loss or cost function (Hastie et al., 2009). Essentially, our goal is to identify the set of weights or parameter values that would yield the smallest possible values in the loss function. The minimization of loss function is the fundamental concept to build neural network models.

The above example we have discussed is for an arbitrary function. Now, connecting this back to our house price prediction example, the loss function will look like this:

$$L(w) = \frac{1}{m}\sum_{i=1}^{m}(y_i - w^T x_i)^2$$

where $L(w)$ is the loss function, y_i is the actual house price, and $\widehat{y}_i$ is the predicted house price, which can be expressed as $\widehat{y}_i = w^T x_i$, where x_i represents the features of the i th house and w represents the weights.

This equation calculates the average squared difference between the actual and predicted house prices (Seber & Lee, 2003), helping to minimize the prediction error. The above loss function is differentiable, which means we can determine the parameter values at which the function reaches its minimum. These parameters at minimum loss are what we are after. This is where gradient descent excels—finding the minima of any differentiable function (Ruder, 2016). In a neural network, we define the loss function as a differentiable function and then pass it to the gradient descent algorithm.

2.2 Cost Function and Gradient Descent

We now have a good intuition of what gradient descent does. Let us now grasp the concept of gradient descent in relation to training neural networks. We will be using logistic regression as an example. In fact, the foundation of neural networks is tied to logistic regression (Cox, 1958). In a neural network, each neuron performs a computation similar to logistic regression. The neuron takes a set of inputs, applies a linear transformation, and then applies a nonlinear function, often called an activation function (Haykin, 1994). This process is similar to logistic regression, where inputs are linearly combined and then transformed using the logistic function. In a neural network, this process is iteratively performed across multiple nodes and layers, enabling the network to learn of complex patterns.

This is why we will start with modeling a logistic regression, which can be related to the modeling of a single node (also known as neuron). In other words, logistic regression can be considered a very simple neural network.

In logistic regression, given x, our goal is to predict $\widehat{y}$. For one example $x^{(i)}$:

$$z^{(i)} = w^T x^{(i)} + b$$

$$\widehat{y}^{(i)} = a^{(i)} = \text{sigmoid}\left(z^{(i)}\right)$$

Note that we are using z and w in lowercase in the notation. This is because they denote quantities that may be either scalars or vectors. Generally speaking, we denote a variable in lowercase if it represents a scalar or a vector. If it is a matrix, we denote the variable with an uppercase letter.

The *sigmoid*(z) is defined as (Press et al., 1992):

$$\sigma(z) = \frac{1}{1 + e^{-z}}$$

The computation of $\widehat{y}$ given x is also known as forward propagation in neural network context. The loss function for logistic regression is defined as:

$$\mathcal{L}\left(a^{(i)}, y^{(i)}\right) = -y^{(i)} \log\left(a^{(i)}\right) - \left(1 - y^{(i)}\right) \log\left(1 - a^{(i)}\right)$$

The cost is then computed by taking average over all training examples:

$$J = \frac{1}{m} \sum_{i=1}^{m} \mathcal{L}\left(a^{(i)}, y^{(i)}\right)$$

Therefore, we want to find parameters w and b by minimizing the cost function $J(w, b)$. This cost minimization is achieved through gradient descent algorithm. This basically means that we want to find parameters w and b that corresponds to minimum of $J(w, b)$. This can be visualized in Fig. 2.2, where the optimization goal is to find minimum of $J(w, b)$.

We can relate this to our house price prediction example. By using logistic regression, we can frame this as a house price classification problem. For instance, we might classify whether a house price falls into the luxury category or if it is a starter home. The parameters w and b can be thought of as the size of a house and the intercept parameter, respectively. Our goal is to find the parameters w and b that have the minimum cost, representing the valley or lowest value in the figure.

It is important to clarify a common point of confusion in gradient descent: the difference between minimizing the loss function and the cost function. The term loss function is employed when we consider a single instance or sample. On the other hand, the term cost function is used when we consider the entire training set. These two concepts are interconnected. Essentially, the cost is the average of the losses computed from all the training samples.

Table 2.1 can help us understand this. Here, we see examples with three data points. For each data point, the loss is calculated using the loss function for logistic regression, the binary cross entropy loss (Murphy, 2012). The cost function is then the average of these losses. This demonstrates how the cost function measures the overall performance of the model across the entire training set, while the loss function provides insight into individual data points.

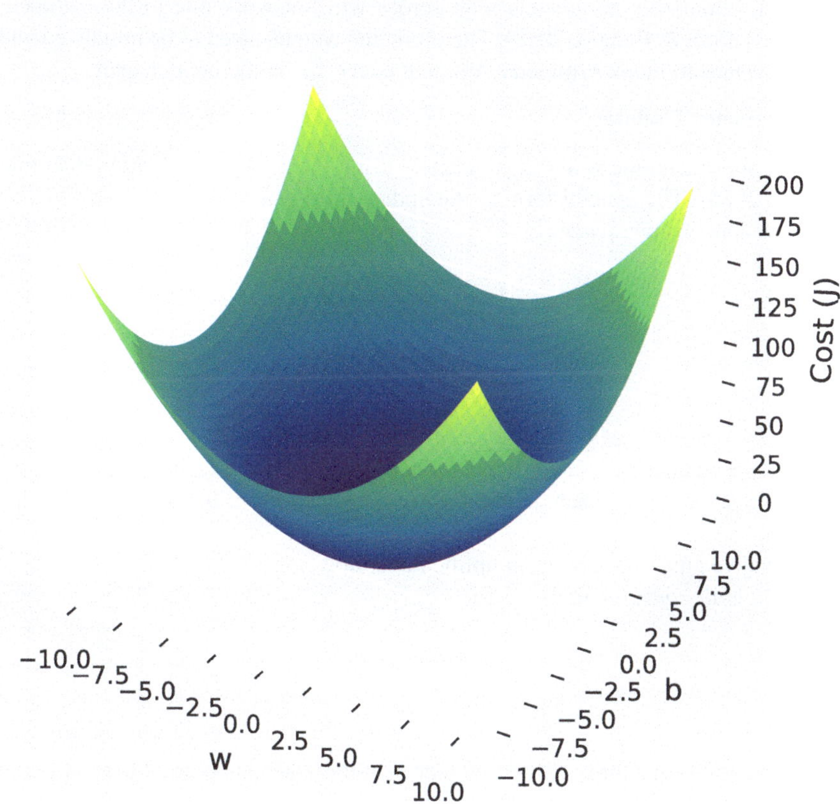

Fig. 2.2 Illustration of finding optimal parameters through cost minimization using gradient descent

Table 2.1 Examples of loss and cost calculations for three data points

Data point	Actual label (y)	Predicted probability (y_hat)	Loss = −[y log(y_hat) + (1 - y) log(1—y_hat)]	Cost = AVG (Loss)
1	1	0.9	0.1054	0.1839
2	0	0.2	0.2231	
3	1	0.8	0.2231	

2.3 Step-by-Step Gradient Descent Algorithm

Here is a step-by-step guide to implement the gradient descent algorithm:

1. **Parameter Initialization:**

We start by initializing the parameters w and b with random values (He et al., 2015). For logistic regression, it is acceptable to initialize parameters with zeroes because the cost function of logistic regression is convex. However, for neural networks, initializing parameters with zeroes will not work due to the symmetry problem (Glorot & Bengio, 2010). Therefore, the weights need to be initialized with random values to break symmetry. We will cover this in the next chapter.

2. **Parameter Update:**

Next, we update the w and b parameters using the gradient descent update rules (Bottou, 2010). The general form of the update rules is as follows:

Iterate until convergence:

$$\left\{ w := w - \alpha \frac{dJ(w)}{dw.} \right.$$

$$\left. b := b - \alpha \frac{dJ(w)}{db} \right\}$$

Here, α is the learning rate, a hyperparameter (Bengio, 2012) that dictates the size of the step taken during each iteration as we navigate toward the minimum of a loss function.

The above equations can be simplified as follows:

Iterate until convergence:

$$\{ w := w - \alpha * dw$$

$$b := b - \alpha * db \}$$

In the above equations, dw and db represent the gradients (partial derivatives) of the cost function with respect to the parameters w and b, respectively. This indicates the change in the cost function with respect to the parameters. This means that gradients tell us how much the loss changes if we increase the parameters by a small amount. They help us understand which parameter affects the loss and by how much.

Note that,

- *Positive gradient*: If the gradient is positive, it means that increasing the parameter will increase the cost.
- *Negative gradient*: If the gradient is negative, it means that increasing the parameter will decrease the cost.

2.4 Non-vectorized Implementation of Gradient Descent

In this section, we will implement a non-vectorized version (Harris et al., 2020) of gradient descent algorithm for logistic regression for a single iteration/step. This will help us to better understand the fundamental mechanics of gradient descent by individually updating each parameter, rather than using vectorized operations. By doing so, we can grasp the computational complexity and the step-by-step process involved in an optimization algorithm. This foundational knowledge will help us to develop intuition before transitioning to more efficient, vectorized implementations.

Let us assume we have two features x_1 and x_2. The forward propagation pass (LeCun et al., 2002) for one training example will be:

$$z = w_1x_1 + w_2x_2 + b$$

and

$$\widehat{y} = \sigma(z)$$

We also know the loss function for logistic regression as discussed in an earlier section. So, our goal is to find the optimal parameters w_1, w_2, and b using the gradient descent algorithm. From our step-by-step gradient descent algorithm, we know that we iteratively adjust the parameters to minimize the loss function. During each iteration, the adjustment is made by subtracting the learning rate multiplied with the gradient of the loss function with respect to the parameter from the current parameter value. Now, the question is how do we compute the gradient of the loss function?

This is easy if we have the partial derivative of the loss function readily available. Let us revisit our arbitrary function. The minimum value of the function $y = (x - 2)^2 + 3$ can be obtained through gradient descent with the following implementation. For this particular function, we also know the derivative of the function, which is $2 * (x - 2)$, thus can update the parameter x by subtracting the derivative of function with respect to the parameter from the current parameter value:

Listing 2.1 Gradient Descent for the Arbitrary Function

```
def func(x):
    return (x-2)**2 + 3

def deriv_func(x):
    return 2*(x-2)

def gradient_descent_func(learning_rate, n_iter):
    x = 100.0
    for i in range(n_iter):
        grad = deriv_func(x)
        x = x - learning_rate * grad
        if i%10 == 0:
            print(f"Iter {i+1}: x = {x}, f(x) = {func(x)}")
    return x
```

Let us now run gradient descent for 50 iterations:

```
learning_rate = 0.1
n_iter = 50

min_value = gradient_descent_func(learning_rate, n_iter)
print(f"\nThe minimum value of the function is at x = {min_value}")
```

This will result in that the minimum value of the function is at $x = 2$:

```
Iter 1: x = 80.4, f(x) = 6149.560000000001
Iter 11: x = 10.41813590016, f(x) = 73.8650120335626
Iter 21: x = 2.9038904596117683, f(x) = 3.8170179629771734
Iter 31: x = 2.0970544990799738, f(x) = 3.0094195757916644
Iter 41: x = 2.0104211474869538, f(x) = 3.000108600314945

The minimum value of the function is at x = 2.0013987027388516
```

Now, going back to the logistic regression, we do not have the derivative of the loss function readily available as we have had for our arbitrary function. This is because there are many operations involved. Therefore, we compute the derivative as part of the learning process using chain rule. The process of computing the derivative of the loss function with respect to the weights and biases (parameters) is done through a method known as backpropagation. This involves performing a forward pass through the network (logistic regression in this case) to compute the output and the loss, and then a backward pass (Werbos, 1994) to compute the gradients. The computations of the forward and backpropagation passes for one training example in the logistic regression example are illustrated in Fig. 2.3.

Referring to our gradient descent step-by-step guide, we will require $\frac{dL}{dw}$ (in short dw) to update parameter w. Let us look at the forward and backward passes in detail:

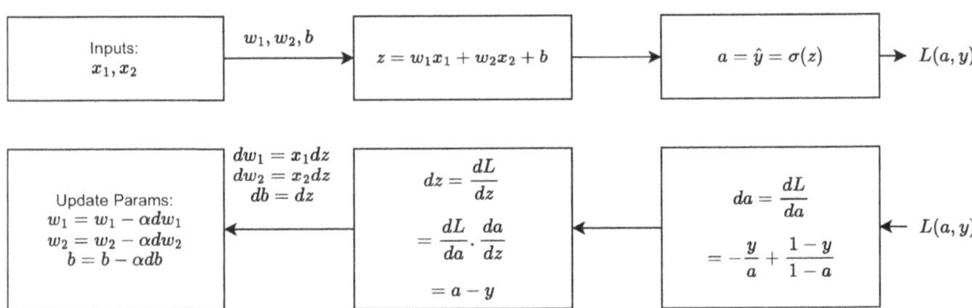

Fig. 2.3 Forward and backward propagations of logistic regression (non-vectorized)

Forward Pass

Operation	Formula
Weighted sum	z = w1*x1 + w2*x2 + b
Activation	a = sigma(z)
Loss	L = −[y*log(a) + (1-y)*log(1-a)]

Backward Pass

First, let us define the derivatives for each individual operation:

Operation	Formula
Loss	dL/da = −(y/a) + ((1-y)/(1-a))
Activation	da/dz = a*(1-a)
Weighted sum	dz/dw1 = x1, dz/dw2 = x2, dz/db = 1

To calculate the gradients for the weights and bias, we can then apply the chain rule. This is because composition of differential function is differentiable as well:

- `dL/dw1 = dL/da * da/dz * dz/dw1`
- `dL/dw2 = dL/da * da/dz * dz/dw2`
- `dL/db = dL/da * da/dz * dz/db`

Note that, we can also do the calculation step by step, from right to left, as shown in Fig. 2.3:

- `Calculate dL/da first, then dL/dz = dL/da * da/dz, then dL/dw1 = dL/dz * dz/dw1`
- `Calculate dL/da first, then dL/dz = dL/da * da/dz, then dL/dw2 = dL/dz * dz/dw2`
- `Calculate dL/da first, then dL/dz = dL/da * da/dz, then dL/db = dL/dz * dz/db`

Substituting the values from the table above:

- `dL/dw1 = -(y/a) + ((1-y)/(1-a)) * a*(1-a) * x1`
- `dL/dw2 = -(y/a) + ((1-y)/(1-a)) * a*(1-a) * x2`
- `dL/db = -(y/a) + ((1-y)/(1-a)) * a*(1-a) * 1`

Finally, the weights are updated using gradient descent as follows:

- `w1 = w1 – alpha * dL/dw1`
- `w2 = w2 – alpha * dL/dw2`
- `b = b – alpha * dL/db`

The following code demonstrates a non-vectorized implementation of the gradient descent algorithm. It performs a single pass to update the parameters `w1`, `w2`, and `b`:

Listing 2.2 Non-vectorized Gradient Descent

```
import numpy as np

def sigmoid(z):
    return 1 / (1 + np.exp(-z))

def gradient_descent_nonvectorized(x1, x2, y):

    J = 0
    dw1 = 0
    dw2 = 0
    db = 0
    w1 = 0
    w2 = 0
    b = 0
    m = len(x1)
    alpha = 0.01

    for i in range(m):
        z = w1*x1[i] + w2*x2[i] + b
        a = sigmoid(z)
        J += y[i]*np.log(a) + (1-y[i])*np.log(1-a)

        dz = a - y[i]
        dw1 += dz * x1[i]
        dw2 += dz * x2[i]
        db += dz

    J /= m
    dw1 /= m
    dw2 /= m
    db /= m

    w1 = w1 - alpha * dw1
    w2 = w2 - alpha * dw2
    b = b - alpha * db

    return w1, w2, b
```

Here, we are simply following the calculations as we discussed earlier. We are implementing both the forward pass and the backward pass for all the operations involved, and subsequently updating the parameters. We can run the above function with the following code snippet. In this snippet, we are assuming we have two features and 20 examples:

```
Nx = 2
m = 20
x1 = np.random.randn(m)
x2 = np.random.randn(m)
y = np.zeros(m)
w1, w2, b = gradient_descent_nonvectorized(x1, x2, y)
```

That is it. This is how we implement gradient descent for a single step or pass. However, we can see that the non-vectorized implementation loops over m examples, which is not efficient. This can be simplified through a vectorized implementation, which we will discuss in the next section.

2.5 Vectorized Implementation of Gradient Descent

Now that we have covered the non-vectorized implementation of gradient descent, let us walk through the vectorized implementation. We will now implement a vectorized version (Harris et al., 2020; Van Der Walt et al., 2011) of gradient descent algorithm for logistic regression for m training examples for a single iteration/step. Figure 2.4 illustrates the forward and backward propagations of logistic regression for a single pass. In vectorized implementation, we are given a feature matrix X with m examples. In one pass, we can compute the activation output for all examples as:

$$A = \widehat{Y} = \sigma\left(w^T X + b\right)$$

The cross entropy cost function is then defined as:

$$J = -\frac{1}{m}\sum\nolimits_{i=1}^{m} y^{(i)} \log\left(a^{(i)}\right) + \left(1 - y^{(i)}\right) \log\left(1 - a^{(i)}\right)$$

For the above forward propagation, the backward propagation (Werbos, 1994) equations will then look like this after applying the chain rule:

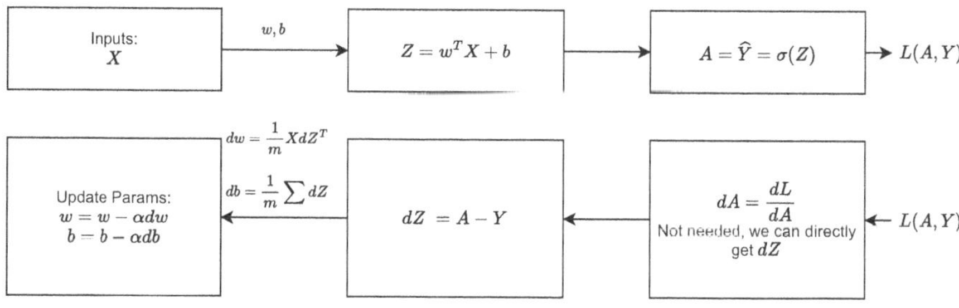

Fig. 2.4 Forward and backward propagations of logistic regression (vectorized)

$$\frac{\partial J}{\partial w} = \frac{1}{m} X(A - Y)^T$$

$$\frac{\partial J}{\partial b} = \frac{1}{m} \sum_{i=1}^{m} \left(a^{(i)} - y^{(i)}\right)$$

Following is the code to implement a vectorized version of the gradient descent algorithm for logistic regression from scratch:

Listing 2.3 Vectorized Gradient Descent

```
def gradient_descent_vectorized(X, Y):
    Nx = X.shape[0]
    w = np.zeros((Nx, 1))
    b = 0

    Z = np.dot(w.T, X) + b
    A = 1 / (1 + np.exp(-Z))
    J = -1/m * np.sum(Y * np.log(A) + (1 - Y) * np.log(1 - A))

    dz = A - Y
    dw = np.dot(X, dz.T) / m
    db = np.sum(dz) / m

    w = w - alpha * dw
    b = b - alpha * db

    return w, b
```

The following code demonstrates how to call the above function. In this example, we assume there are 4 features and 500 training examples. This means the rank of `X` matrix is 2 (i.e., it is a two-dimensional matrix). We are using learning rate of 0.01 to calculate the parameters w and b for a single step:

```
Nx = 4
m = 500
X = np.random.randn(Nx, m)
Y = np.zeros([1, m])
alpha = 0.01
w, b = gradient_descent_vectorized(X, Y)
```

That is it! This is the gradient descent implementation from scratch, which is more efficient than the non-vectorized implementation. With the vectorized approach, we do not have to loop over m examples to calculate gradients and update parameters. We can rather do this efficiently.

2.6 Summary

- Neural networks map feature space to target space using historical observations, and gradient descent helps learn this mapping by finding the minimum of differentiable functions, particularly the loss or cost function.
- Logistic regression serves as a foundation for understanding neural network training, with computations similar to logistic regression performed across multiple nodes and layers in a neural network.
- Gradient descent minimizes the cost function by iteratively adjusting parameters, starting with random initialization to avoid symmetry problems, and updating parameters using learning rate-controlled steps until convergence.
- A non-vectorized implementation of gradient descent for logistic regression involves forward and backward propagation to compute outputs, losses, and gradients, followed by iterative parameter updates to minimize the loss function.
- A vectorized implementation of gradient descent for logistic regression enhances efficiency by using matrix operations for simultaneous computations of outputs, gradients, and parameter updates.

References

Bengio, Y. (2012). Practical recommendations for gradient-based training of deep architectures. In *Neural networks: Tricks of the trade: Second edition* (pp. 437–478). Springer.

Bishop, C. M., & Nasrabadi, N. M. (2006). *Pattern recognition and machine learning* (Vol. 4). Springer.

Bottou, L. (2010). Large-scale machine learning with stochastic gradient descent. *Proceedings of COMPSTAT'2010: 19th International Conference on Computational StatisticsParis France, August 22–27, 2010 Keynote, Invited and Contributed Papers*, 177–186.

Boyd, S. P., & Vandenberghe, L. (2004). *Convex optimization*. Cambridge university press.

Cox, D. R. (1958). The regression analysis of binary sequences. *Journal of the Royal Statistical Society Series B: Statistical Methodology*, *20*(2), 215–232.

Glorot, X., & Bengio, Y. (2010). Understanding the difficulty of training deep feedforward neural networks. *Proceedings of the Thirteenth International Conference on Artificial Intelligence and Statistics*, 249–256.

Goodfellow, I. (2016). *Deep learning*. MIT press.

Harris, C. R., Millman, K. J., Van Der Walt, S. J., Gommers, R., Virtanen, P., Cournapeau, D., Wieser, E., Taylor, J., Berg, S., Smith, N. J., & others. (2020). Array programming with NumPy. *Nature*, *585*(7825), 357–362.

Hastie, T., Tibshirani, R., Friedman, J., & others. (2009). *The elements of statistical learning*. Citeseer.

Haykin, S. (1994). *Neural networks: A comprehensive foundation*. Prentice Hall PTR.

He, K., Zhang, X., Ren, S., & Sun, J. (2015). Delving deep into rectifiers: Surpassing human-level performance on ImageNet classification. *Proceedings of the IEEE International Conference on Computer Vision*, 1026–1034.

LeCun, Y., Bottou, L., Orr, G. B., & Müller, K.-R. (2002). Efficient backprop. In *Neural networks: Tricks of the trade* (pp. 9–50). Springer.

Murphy, K. P. (2012). *Machine learning: A probabilistic perspective*. MIT press.

Nocedal, J., & Wright, S. J. (1999). *Numerical optimization*. Springer.

Press, W. H., Teukolsky, S. A., Vetterling, W. T., & Flannery, B. P. (1992). *Numerical Recipes (Cambridge*. Cambridge Univ. Press.

Ruder, S. (2016). An overview of gradient descent optimization algorithms. *arXiv Preprint arXiv:1609.04747*.

Rumelhart, D. E., Hinton, G. E., & Williams, R. J. (1986). Learning representations by back-propagating errors. *Nature*, *323*(6088), 533–536.

Seber, G. A., & Lee, A. J. (2003). *Linear regression analysis*. John Wiley & Sons.

Van Der Walt, S., Colbert, S. C., & Varoquaux, G. (2011). The NumPy array: A structure for efficient numerical computation. *Computing in Science & Engineering*, *13*(2), 22–30.

Werbos, P. J. (1994). *The roots of backpropagation: From ordered derivatives to neural networks and political forecasting*. John Wiley & Sons.

Chapter 3
Training Deep Neural Networks

This chapter covers:

- Understanding the fundamental components of neural networks.
- Exploring the architecture of neural networks.
- Discussing activation functions and matrix dimensions.
- Detailing the training and prediction steps involved in training a neural network.
- Providing a practical example of credit classification.

In the previous chapter, we explored the workings of the gradient descent algorithm. Specifically, we learned that gradient descent minimizes the cost function by iteratively updating parameters. This minimization process, in fact, determines the optimal parameters. However, so far, we have discussed implementing a gradient descent algorithm on a single logistic regression node, not on a full neural network. This chapter builds upon our understanding of gradient descent on single neuron and expands to the creation of a full neural network, capable of complex learning through interconnected layers. We will explore the integration of data, layers, loss functions, and optimization algorithms in the training process of deep neural networks. We will also apply this knowledge to a practical example: credit classification.

3.1 Neural Network Components

Now we have a basic understanding of the gradient descent algorithm. We can thus proceed to the training process of a neural network using the gradient descent algorithm (Ruder 2016). The process of training a neural network is fundamentally structured around four core components:

- Data.
- Layers and architecture.

T. Islam, *Hands-on Deep Learning*, https://doi.org/10.1007/978-3-032-00488-8_3

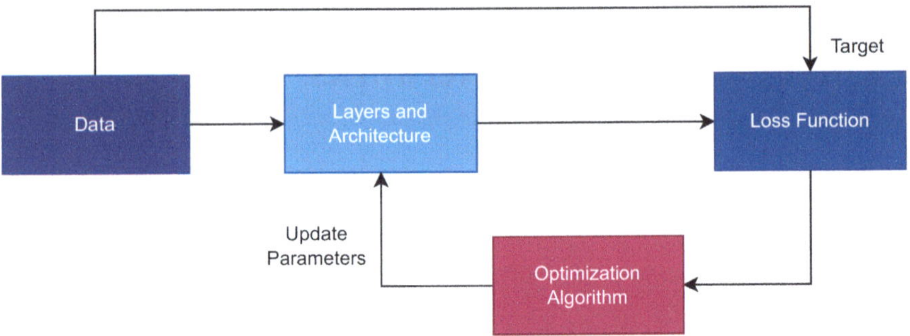

Fig. 3.1 This figure outlines the key components involved in training a neural network. It includes elements such as the data, layers and architecture, loss function, and optimization algorithm, illustrating how network learns and updates the parameters

- Loss function.
- Optimization algorithm.

These components (Fig. 3.1) are interconnected, forming the backbone of a neural network (Goodfellow 2016). The absence of even a single component can prevent the network from training at all. Data is the lifeblood of neural networks (Halevy et al. 2009). Just as we learn from experience, neural networks require a substantial dataset to identify patterns and make accurate predictions. Neural network layers and architecture define the network's structure (LeCun et al. 2015), outlining how information flows through the system. It dictates the complexity of the model and its ability to capture complex relationships within the data. Loss function acts as the performance evaluator, measuring how well the network's predictions align with the actual outcomes. This evaluation guides the learning process, allowing the network to adjust its internal parameters. Optimization algorithm iteratively tweaks the network's internal parameters (weights and biases) to minimize the loss function. Through this process, the network progressively improves its performance.

In the previous chapter, we already covered into the workings of optimization algorithms, with a particular focus on gradient descent. Now, let us shift our attention to the layers and architecture of a fully connected neural network, often referred to as dense neural network.

3.2 Layers and Architecture

In the context of a fully connected neural network, the architecture consists of three types of layers (Schmidhuber 2015):

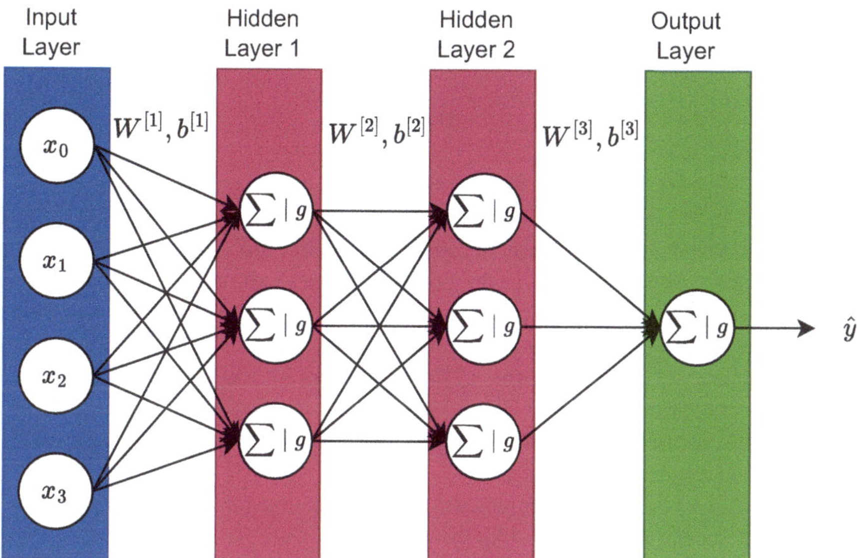

Fig. 3.2 A three-layer neural network architecture. It includes the input layer, hidden layers, and output layer, along with the connections between them

- *Input layer*: This layer receives the input data.
- *Hidden layers*: These layers follow the input layer. Neural networks may have as few as one or two hidden layers, while larger networks can have hundreds (deep neural networks).
- *Output layer*: This is the final layer, producing the results for the given inputs.

Let us examine the neural network architecture depicted in Fig. 3.2. This diagram illustrates a three-layer neural network, which includes one input layer, two hidden layers, and an output layer. One may ask why it is referred to as a three-layer network when there appear to be four layers in the figure? The reason for this is that the input layer, while crucial, is not typically counted in the layer tally of a neural network.

In a hidden layer, there can be many hidden neurons. These neurons are occasionally referred to as nodes in the literature as well. The computational process of an individual node is akin to the calculations performed in logistic regression as we covered in the previous chapter. Recall from the previous chapter, a single node's activation is defined as:

$$z^{(i)} = w^T x^{(i)} + b$$

$$a^{(i)} = g\left(z^{(i)}\right)$$

where g is the activation function.

The above calculation can be expanded for a layer consisting of many hidden nodes:

$$z^{(i)} = Wx^{(i)} + b$$

$$a^{(i)} = g\left(z^{(i)}\right)$$

Note that W is in uppercase here, because it is a matrix that represents the weights of connections from all neurons in the previous layer to all neurons in the current layer. Hence, it is a two-dimensional structure, i.e., a matrix. The b is in lowercase, because it is a vector that represents the bias for each neuron in the current layer. Each neuron has only one bias term, so for all neurons in a layer, the biases form a one-dimensional structure, i.e., a vector. Since we are discussing a single example, we use z and a, as they are vectors. However, for m examples, we would use Z and A. We will see examples of this later.

With all this, we can perform forward pass calculation for all m examples. The Python implementation will look like this:

```
for i in range(m):
    z1 = np.dot(W1, x) + b1
    a1 = relu(z1)
    z2 = np.dot(W2, a1) + b2
    a2 = relu(z2)
    z3 = np.dot(W3, a2) + b3
    a3 = sigmoid(z3)
```

It is worth noting that g is the activation function. There are variants of activation functions, such as relu and sigmoid. They will be discussed in the next section.

3.3 Activation Functions

Activation functions introduce nonlinearity into the learning process of a neural network (Nair and Hinton 2010). Because of this nonlinearity, neural networks are very powerful in learning complex patterns from data. When relating this to a biological neural network, we can think of activation functions as determining whether a particular neuron should be activated (fired) or not. Activation functions in a neural network need to be differentiable for backpropagation to work. Table 3.1 provides some of the most common activation functions used in neural networks.

For the sake of reference, let us implement activation functions based on Table 3.1. In the latter part of the chapter, we will see an example of the use of this in a case study:

Table 3.1 Common activation functions used in neural networks

Activation function	Equation	Derivative
Sigmoid	$a = \frac{1}{1+e^{-z}}$	$a(1 - a)$
Tanh	$a = \frac{e^z - e^{-z}}{e^z + e^{-z}}$	$1 - a^2$
ReLU	$a = \max(0, z)$	$\begin{cases} 0 & \text{if } z < 0 \\ 1 & \text{if } z \geq 0 \end{cases}$
Leaky ReLU	$a = max(0.01z, z)$	$\begin{cases} 0.01 & \text{if } z < 0 \\ 1 & \text{if } z \geq 0 \end{cases}$

Listing 3.1 Most Common Activation Functions

```
def relu(Z):
    return np.maximum(0, Z)

def sigmoid(Z):
    return 1 / (1 + np.exp(-Z))

def tanh(Z):
    return np.tanh(Z)

def leaky_relu(Z):
    return np.maximum(0.01, Z)
```

Their derivatives will be:

Listing 3.2 Derivatives of the Activation Functions

```
def relu_derivative(Z):
    return np.where(Z < 0, 0, 1)

def sigmoid_derivative(Z):
    A = sigmoid(Z)
    return A * (1 - A)

def tanh_derivative(Z):
    A = tanh(Z)
    return 1 - A**2

def leaky_relu_derivative(Z):
    return np.where(Z < 0, 0.01, 1)
```

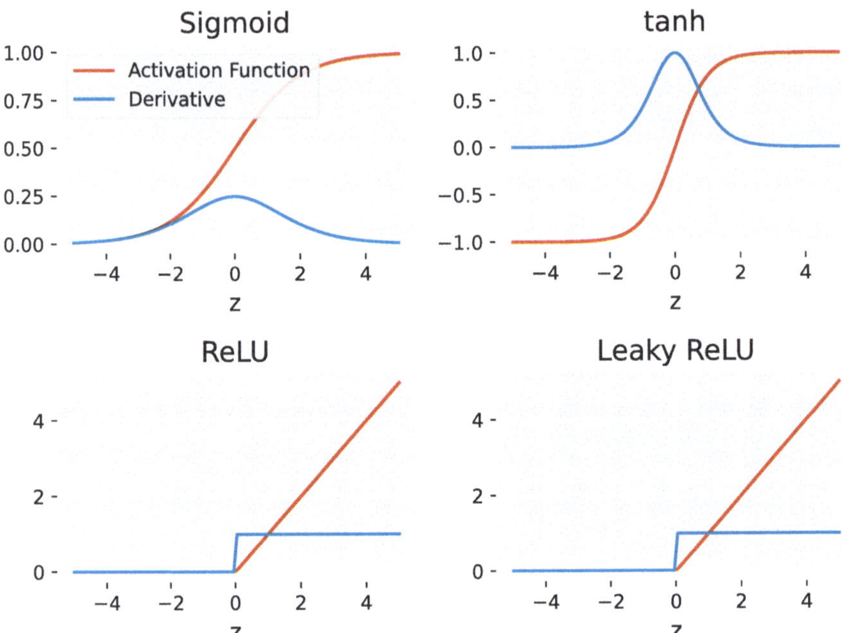

Fig. 3.3 Some of the most commonly used activation functions in neural networks. It also illustrates the derivatives of these functions, which are essential for the backpropagation process in training neural networks

In Fig. 3.3, we illustrate the properties of various activation functions. We can see that the derivatives of both the sigmoid and tanh activation functions approach zero near their extremes. This characteristic poses a challenge for neural network training because when the derivatives are zero, the parameters remain unchanged. This phenomenon is known as the *vanishing gradient problem*, as the gradients effectively disappear. Therefore, the sigmoid activation function is never used in hidden layers due to its susceptibility to the vanishing gradient problem. Instead, it is used in the output layer for binary classification tasks. The tanh function also suffers from the same issue and not commonly used these days. The Rectified Linear Unit (ReLU) activation function (He et al. 2015; Nair and Hinton 2010) addresses this problem to a certain extent, as its derivative is not zero for positive values. Therefore, ReLU is a popular choice of activation function for hidden layers. The Leaky ReLU is a slight modification of the ReLU activation function (Maas et al. 2013), designed to prevent the activation output from reaching exactly zero. This adjustment helps mitigate the vanishing gradient problem, making Leaky ReLU another viable option for hidden layers.

3.4 Matrix Dimensions and Parameters

Now that we have covered the activation functions, let us look at the matrix dimensions for our three-layer neural networks. Our input layer contains the feature matrix X. In first hidden layer, we compute $Z^{[1]}$ matrix, followed by activation output matrix $A^{[1]}$. In second hidden layer, we compute $Z^{[2]}$ matrix, followed by activation output matrix $A^{[2]}$. Finally, in output layer, we compute $Z^{[3]}$ matrix, followed by activation output matrix $A^{[3]}$ that is also the prediction matrix. Table 3.2 tabulates the dimensions in detail.

Now, we know the dimensions for inputs and operations for the three-layer neural networks. In addition to this, it is important to know the dimensions of the weight matrices and bias vectors as well. This is because we have to ensure that the dimensions of the weight matrices and bias vectors match the input and output dimensions of each layer for the neural network to function correctly. Mismatched dimensions can lead to errors during the training process. The dimensions of the weight matrices and bias vectors will also determine the total number of parameters in the model. Next, the dimensions of the weight matrices and bias vectors for the three layers are provided in Table 3.3.

Table 3.2 Matrix dimensions for neural network inputs and operations

Matrix	Dimension
X	This is the input matrix with dimensions **(nx, m)**, where **nx** is the number of input features and **m** is the number of training examples
$Z^{[1]}$	This is the pre-activation weighted sum matrix for the first hidden layer. It has dimensions **(n1, m)**, where **n1** is the number of units in the first hidden layer and **m** is the number of training examples
$A^{[1]}$	This is the activation matrix for the first hidden layer. It also has dimensions **(n1, m)**, where **n1** is the number of units in the first hidden layer and **m** is the number of training examples. This matrix is obtained by applying the activation function to the pre-activation weighted sum matrix $Z^{[1]}$
$Z^{[2]}$	This is the pre-activation weighted sum matrix for the second hidden layer. It has dimensions **(n2, m)**, where **n2** is the number of units in the second hidden layer and **m** is the number of training examples
$A^{[2]}$	This is the activation matrix for the second hidden layer. It also has dimensions **(n2, m)**, where **n2** is the number of units in the second hidden layer and **m** is the number of training examples. This matrix is obtained by applying the activation function to the pre-activation weighted sum matrix $Z^{[2]}$
$Z^{[3]}$	This is the pre-activation weighted sum matrix for the output layer. It has dimensions **(ny, m)**, where **ny** is the number of units in the output layer and **m** is the number of training examples
$A^{[3]}$ or $\widehat{y}$	This is the activation matrix for the output layer. It also has dimensions **(ny, m)**, where **ny** is the number of units in the output layer and **m** is the number of training examples. This matrix is obtained by applying the activation function to the pre-activation weighted sum matrix $Z^{[3]}$. The activation function used here is typically the softmax function for multi-class classification problems, or the sigmoid function for binary classification problems

Table 3.3 The dimensions of the weight matrices and bias vectors for the three layer neural network

Weight matrix and bias vector	Dimension
$W^{[1]}$	This is the weight matrix for the first hidden layer. It has dimensions **(n1, nx)**, where **n1** is the number of units in the first hidden layer and **nx** is the number of input features
$b^{[1]}$	This is the bias vector for the first hidden layer. It has dimensions **(n1, 1)**
$W^{[2]}$	This is the weight matrix for the second hidden layer. It has dimensions **(n2, n1)**, where **n2** is the number of units in the second hidden layer and **n1** is the number of units in the first hidden layer (previous layer)
$b^{[2]}$	This is the **bias vector** for the second hidden layer. It has dimensions **(n2, 1)**
$W^{[3]}$	This is the weight matrix for the output layer. It has dimensions **(ny, n2)**, where **ny** is the number of units in the output layer and **n2** is the number of units in the second hidden layer (previous layer)
$b^{[3]}$	This is the bias vector for the output layer. It has dimensions **(ny, 1)**

Table 3.4 The dimensions of the weight matrices and bias vectors for the L layer neural network

Layer l	Weight matrix $W^{[l]}$	Bias vector $b^{[l]}$	Pre-activation matrix $Z^{[l]}$	Activation matrix $A^{[l]}$
l	$(n[l], n[l\text{-}1])$	$(n[l], 1)$	$(n[l], m)$	$(n[l], m)$

The matrix dimensions and parameters provided above correspond to our three-layer neural network architecture. This can be extended to an l-layer neural network architecture, as shown in Table 3.4.

In this table,

- l is the layer number, ranging from 1 (the first hidden layer) to L (the output layer).
- $n[l]$ is the number of units in layer l.
- $n[l\text{-}1]$ is the number of units in the previous layer (l-1).
- m is the number of training examples.
- $W[l]$ is the weight matrix for layer l, with dimensions $(n[l], n[l-1])$.
- $b[l]$ is the bias vector for layer l, with dimensions $(n[l], 1)$.
- $A[l]$ is the activation matrix for layer l, with dimensions $(n[l]$, m).
- $Z[l]$ is the pre-activation matrix for layer l, with dimensions $(n[l], m)$.

This table represents the dimensions of the parameters and variables for any layer in a fully connected deep neural network. The weights and biases (W and b) are parameters that the network learns during training. The activation and pre-activation matrices (A and Z) are calculated during the forward propagation step. The dimensions of these matrices and vectors depend on the number of units in each layer and the number of training examples. The process of learning the parameters and calculating these matrices is repeated for each layer in the network, from the first hidden layer to the output layer. This is how the network is able to learn complex patterns from the input data and make accurate predictions.

3.5 Neural Network Training and Prediction Steps

With the basics out of the way, let us now move on to the training and prediction steps. Figure 3.4 depicts the steps of building a neural network from scratch. We start by normalizing the input features to ensure that the data is on a similar scale. Next, we initialize the parameters, which are the weights and biases of the network. After that, we define the loss function, which measures how well the network's predictions match the actual results. The core of the process involves looping for a number of iterations. During each iteration, we perform forward propagation to compute the network's output, calculate the cost to see how far off the predictions are, carry out backward propagation to adjust the parameters based on the error, and finally update the parameters to improve the network's performance. Once the network is trained, we can use the trained parameters to make predictions on new data.

3.5.1 Credit Classification Problem

We will now dive deep into the training and prediction steps of a neural network using a real-world example. We will be utilizing the "German Credit Numeric" dataset from TensorFlow datasets for this purpose. The objective is to classify individuals into categories of good or bad credit risks based on various attributes such as credit amount, credit history, employment status, and so forth. We will be using the "numeric" variant of the dataset, which means that categorical and ordered categorical attributes have already been encoded for compatibility with a neural network.

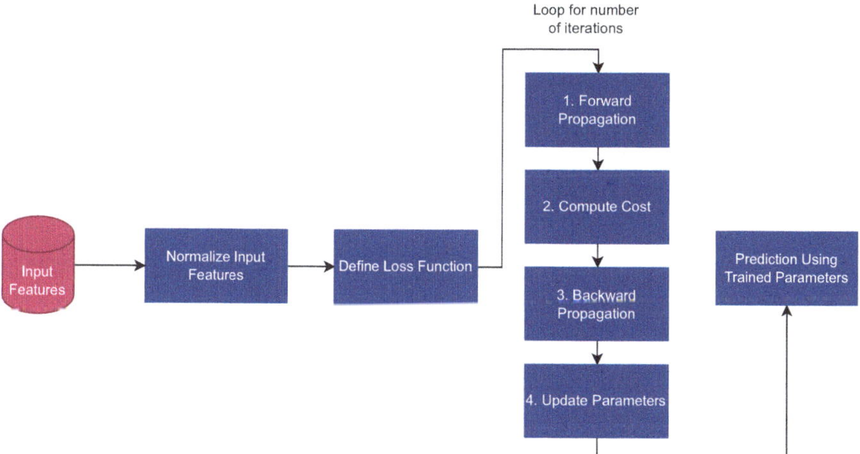

Fig. 3.4 The steps involved in training a neural network, including normalizing input features, defining loss function, model training, and prediction

First things first. Let us first load the German Credit Numeric dataset from TensorFlow dataset. The following code shows how to do this:

Listing 3.3 Loading German Credit Numeric Dataset

```
def load_german_credit_numeric_data():
    dataset, metadata = tfds.load(
        "german_credit_numeric", with_info=True, split="train"
    )
    df = tfds.as_dataframe(dataset)
    X = np.stack(df["features"].values)
    y = df["label"].values
    return X, y
```

With this, we can load and then generate input features and target data:

```
X, y = load_german_credit_numeric_data()
```

Next, we split the dataset into train and test set:

Listing 3.4 Function to Split Dataset into Train and Test Set

```
def split_dataset(X, y, test_size=0.2, random_state=42):
    X_train, X_test, y_train, y_test = train_test_split(
        X, y, test_size=test_size, random_state=random_state
    )
    return X_train, X_test, y_train, y_test
```

Let us apply this function:

```
X_train, X_test, y_train, y_test = split_dataset(X, y)
```

We now have the datasets ready to proceed with the training and prediction steps of a neural network.

3.5.2 *Normalize Input Features*

The initial step in training a neural network involves the normalization of the input feature matrix, which we have denoted as X. Normalization is a crucial technique that standardizes all input features to a consistent scale, typically ranging from 0 to 1 or -1 to 1 or of small values. Consider a scenario where we are predicting house prices based on two distinct features: the size of the home in square kilometers and the number of bedrooms in the house. These features significantly differ in their scales, which can introduce challenges when training a neural network using the original, raw scale. Training on the raw scale can lead to numerical instability during the learning process. The gradient descent algorithm, which is responsible for updating the parameters and facilitating convergence, may struggle due to the elongated shape of the loss function's contour. This elongation can result in instability during the learning process.

Moreover, without normalization, features with larger scales can dominate the learning process. This dominance can lead to a model that is biased toward these larger-scale features, thereby affecting the accuracy and fairness of the predictions. Therefore, normalization is an essential step in preparing data for a neural network, ensuring stability, efficiency, and unbiased learning.

Note that if normalization is applied to the training data, the same fitted parameters must also be applied to the test data during the prediction step.

Following is the code to normalize input features by removing the mean and scaling to unit variance using Sklearn (Pedregosa et al. 2011):

Listing 3.5 Function to Normalize Input Features

```
from sklearn.preprocessing import StandardScaler

def normalize_features(X_train, X_test):

    scaler = StandardScaler()
    X_train = scaler.fit_transform(X_train)
    X_test = scaler.transform(X_test)

    return X_train, X_test
```

Mathematically, this is defined as:

$$x_{scaled} = (x - u)/s$$

In the above equation, μ is the mean of the training samples and s is the standard deviation of the training samples. The distribution of values after applying StandardScaler will have a mean of 0 and a standard deviation of 1.

Next, we apply this feature normalization on our credit card classification train and test set:

```
X_train, X_test = normalize_features(X_train, X_test)
```

3.5.3 Initialize Parameters

The subsequent step in training a neural network involves the initialization of parameters, denoted as W and b. At first glance, one might be inclined to initialize these parameters with zero values. However, this approach will unfortunately not work due to an issue known as symmetry. When all parameters are initialized to zero or to a constant value, it implies that all weights are identical and same throughout the network. As a result, during the forward propagation phase, each neuron in the hidden layer receives identical signals. This uniformity subsequently causes them to follow the same gradient during the backpropagation phase. The consequence of this symmetry is that the model fails to learn anything unique from the data. Each neuron in the hidden layer essentially mimics the others, and the overall network lacks the diversity needed for effective learning. Therefore, it is crucial to break this symmetry when initializing the parameters.

The issue of symmetry can be effectively addressed by initializing parameters randomly (Sutskever et al. 2013). Specifically, the weight parameters, denoted as W, should be initialized with random values. As for the bias parameters, represented as b, initialization with zeros is generally acceptable.

However, it is important to note that while random initialization of parameters may prove effective in certain scenarios, it is not a recommended solution. In some instances, the training process can become suboptimal, and the model's learning may "stuck." This phenomenon is often attributed to the issues of vanishing and exploding gradients, which are particularly prevalent in deep neural networks with numerous layers (Bengio et al. 1994). This is because when neural network is deep, the derivatives can get really big or really small.

Let us look at an example in the following code snippet. Consider the scenario of a deep neural network with 100 layers. As we traverse from the input layer to the output layer during forward propagation, the activation outputs may increase in magnitude. By the time we reach the 100th layer, the activation output has grown exponentially from what it was at the first layer. This phenomenon is not isolated to forward propagation. During backpropagation, when we calculate gradients, a similar pattern may emerge. The gradients can also grow rapidly due to the repeated multiplication of large weight values across layers. This is an example of the "exploding gradients" problem.

```
a = 1
L = 100
w = 1.5
for l in range(1, L+1):
    z = w*a
    a = relu(z)
    if l%20 == 0:
        print(f"Layer: {l}, Activation: {a}")
```

```
Layer: 20, Activation: 3325.256730079651
Layer: 40, Activation: 11057332.320940012
Layer: 60, Activation: 36768468716.93302
Layer: 80, Activation: 122264598055704.66
Layer: 100, Activation: 4.065611775352152e+17
```

In contrast, let us consider a different scenario demonstrated in the following code snippet. This time, we observe the phenomenon of the vanishing gradient. As we progress from the first layer to the 100th layer during forward propagation, the activation outputs decrease exponentially. By the time we reach the 100th layer, the activation output has diminished significantly from its initial value at the first layer. During backpropagation, a similar pattern can emerge, where gradients become smaller as they propagate backward through the network. As we move toward the input layer, these gradients decrease even further, nearly vanishing. This is a classic example of the "vanishing gradient" problem.

```
a = 1
L = 100
w = 0.5
for l in range(1, L+1):
    z = w*a
    a = relu(z)
    if l%20 == 0:
        print(f"Layer: {l}, Activation: {a}")
```

```
Layer: 20, Activation: 9.5367431640625e-07
Layer: 40, Activation: 9.094947017729282e-13
Layer: 60, Activation: 8.673617379884035e-19
Layer: 80, Activation: 8.271806125530277e-25
Layer: 100, Activation: 7.888609052210118e-31
```

So, summarizing, the above phenomenon is known as vanishing and exploding gradients in deep neural networks. This problem often occurs in deep neural networks with many layers, where the magnitude of gradients can decrease (vanish) or increase (explode) exponentially during backpropagation, making the network hard to train.

There are two popular initialization methods that partially address vanishing and exploding gradients problem. They are known as He and Xavier initializations. These initializations draw the weights from a distribution with mean 0 and a defined variance: $2/n$ for He and $1/n$ for Xavier. By initializing the weights in this way, He and Xavier initializations help to balance the variance of the gradients across layers, preventing them from vanishing or exploding too quickly. This leads to a more stable and efficient training of deep neural networks.

He Initialization

He initialization (He et al. 2015) is designed for layers with ReLU activation. It draws weights from a zero-centered Gaussian distribution with a variance (scaling factor) that depends on the nodes in previous layer. The equation is:

$$W^{[l]} = randn\left(\left(n^{[l]}, n^{[l-1]}\right)\right) \times \sqrt{\frac{2}{n^{[l-1]}}}$$

Xavier Initialization

Xavier initialization, also known as Glorot initialization (Glorot and Bengio 2010), is designed for layers with sigmoid or tanh activations. It also draws weights from a zero-centered Gaussian distribution with a variance (scaling factor) that depends on the nodes in previous layer. The equation is:

$$W^{[l]} = randn\left(\left(n^{[l]}, n^{[l-1]}\right)\right) \times \sqrt{\frac{1}{n^{[l-1]}}}$$

Here:

- $W^{[l]}$ is the weight matrix
- $n^{[l-1]}$ is the number of input units in the weight tensor
- $randn((n^{[l]}, n^{[l-1]}))$ is a random matrix drawn from a standard normal distribution.

We now implement these two initialization methods in the following code:

Listing 3.6 Function to Initialize Neural Network Parameters

```
def initialize_parameters(layer_dims, initialization="he"):
    np.random.seed(100)
    parameters = {}
    L = len(layer_dims)-1

    for l in range(1, L+1):
        if initialization == "he":
            parameters['W' + str(l)] = np.random.randn(layer_dims
[l], layer_dims[l - 1]) * np.sqrt(2 / layer_dims[l - 1])
        elif initialization == "xavier":
            parameters['W' + str(l)] = np.random.randn(layer_dims
[l], layer_dims[l - 1]) * np.sqrt(1 / layer_dims[l - 1])
        parameters['b' + str(l)] = np.zeros((layer_dims[l], 1))

    return parameters
```

Therefore, if we want to train a three-layer neural network for our credit classification task, we can initialize the parameters as follows:

```
hidden_layer_sizes = [3, 3]
layer_sizes = [nx] + hidden_layer_sizes + [1]
parameters = initialize_parameters(layer_sizes)
```

3.5.4 *Define Loss Function*

The next step in training a neural network is defining a loss function for computing loss or costs. The loss function quantifies the discrepancy between the network's predictions and the actual targets. The loss function plays a crucial role in training a neural network as it measures the error for the training dataset. The cost value is then used by the optimizer to adjust the network's weights. Table 3.5 lists the popular loss functions used in training deep neural networks.

A binary classification problem has only two classes (yes/no). In contrast, a multi-label classification problem allows for multiple labels to be true simultaneously. For instance, in image classification, an image could be classified as both "dog" and "cat" if both are present in the image. For multi-label classification problems, "binary cross-entropy" is often used as the loss function. In this scenario, each label is considered as a separate binary classification problem and the network predicts whether each label is present or not, independently of the others. Regression problem deals with predicting continuous values.

Table 3.5 Popular loss functions used in training deep neural networks

Loss function	Description	Output layer activation function
Binary cross-entropy	This loss function is used for binary or multi-label classification problems. It calculates the cross-entropy loss between true labels and predicted labels	**Sigmoid**
Categorical cross-entropy	This loss function is used for multi-class classification problems where the labels are one-hot encoded. It calculates the cross-entropy loss between true labels and predicted labels	**Softmax**
Sparse categorical cross-entropy	This loss function is also typically used for multi-class classification problems. It is similar to categorical cross-entropy but is used when the labels are integers instead of one-hot encoded. It calculates the cross-entropy loss between true labels and predicted labels	**Softmax**
Mean squared error (MSE)	This loss function is used for regression problems. It calculates the average squared difference between the true and predicted labels	**Linear (no activation)**

Here are the formulas for the aforementioned loss functions:

1. Binary cross-entropy

$$J = -\frac{1}{m}\sum_{i=1}^{m}\left(y^{(i)}\log\left(a^{[L](i)}\right) + \left(1 - y^{(i)}\right)\log\left(1 - a^{[L](i)}\right)\right)$$

2. Categorical cross-entropy

$$J = -\frac{1}{m}\sum_{i=1}^{m}\sum_{j=1}^{n} y_j^{(i)}\log\left(a_j^{[L](i)}\right)$$

where n is the number of classes.

3. Sparse categorical cross-entropy

$$J = -\frac{1}{m}\sum_{i=1}^{m}\log(a_{y^{(i)}}^{[L](i)})$$

where $y^{(i)}$ is the true class index for the "ith" example.

4. Mean squared error (MSE)

$$J = \frac{1}{m} \sum_{i=1}^{m} \left(y^{(i)} - a^{[L](i)} \right)^2$$

In these equations, m is the number of examples, $y^{(i)}$ is the true label for the "ith" example, and $a^{[L](i)}$ is the predicted label for the "ith" example. The superscript "$[L]$" denotes the output layer of the neural network.

Since our credit classification problem is a binary classification problem, we will need to apply the binary cross-entropy loss function (Hand and Henley 1997). We can implement the computation of the loss (non-vectorized) for the binary cross-entropy loss function as follows:

Listing 3.7 Binary Cross-Entropy Loss Function

```
def compute_loss(y_hat, y):

    loss = - (y*np.log(y_hat) + (1-y)*np.log(1-y_hat))

    return loss
```

The following code expands the computation of loss into the computation of cost (vectorized):

Listing 3.8 Computing Binary Cross-Entropy Cost

```
def compute_cost(Y_hat, Y):

    m = Y.shape[1]
    cost = - (1 / m) * np.sum(Y * np.log(Y_hat) + (1 - Y) * np.log(1 -
Y_hat))

    return cost
```

3.5.5 *Forward and Backward Propagation*

The next step in training a neural network involves a process that consists of both forward and backward propagation (LeCun et al. 2002; Werbos 1994). We also update the parameters iteratively.

Non-vectorized Training

A non-vectorized version of forward and backward propagation formulas for our three-layer neural network for *i*th example are shown in Fig. 3.5. In forward propagation, as depicted in the top panel, the input data is passed through the network layer by layer. At each layer, pre-activation weighted sum is computed, which is then transformed by an activation function to produce the activation value. This process is repeated for each layer, propagating the information forward through the network. The final output is a prediction based on the input data.

Backward propagation, shown in the bottom panel of the figure, is the process by which the network learns from its errors. It starts with the calculation of the loss, which measures the discrepancy between the network's prediction and the actual target. The computation then involves determining the partial derivative of the loss function for each weight within the network. This is where the chain rule comes into play. The chain rule (Spivak 1994) enables us to calculate the derivative of the loss function concerning its inputs. This is particularly useful when the function consists of several nested functions, as seen in a neural network. We should have already developed an intuition, as we discussed this in detail in the previous chapter. In the context of backward propagation, we are extending the concept to multiple layers. The chain rule is used to "chain" the gradients from the output layer back to the input layer, hence the name "backward propagation."

This gradient information represents how much a small change in each weight would affect the final loss. The weights are then updated in the direction that reduces the loss, typically by subtracting the gradient scaled by a learning rate. This process of forward propagation, backward propagation, and weight update is repeated iteratively until the network's predictions are satisfactory.

The following code implements a non-vectorized version of training that consists of forward and backward propagations from scratch:

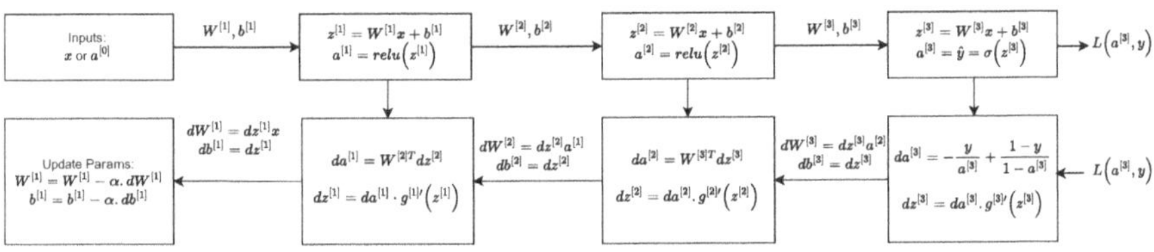

Fig. 3.5 The non-vectorized version of forward and backward propagation formulas for the three-layer neural network. It details the step-by-step calculations involved in each layer

Listing 3.9 Non-vectorized Training of a Neural Network Model

```
def train_nonvec(X, y, learning_rate, num_iterations, hidden_
layer_sizes):

    X = X.T
    Y = y.reshape(1, -1)
    nx, m = X.shape

    layer_sizes = [nx] + hidden_layer_sizes + [1]
    parameters = initialize_parameters(layer_sizes)
    W1 = parameters['W1']
    b1 = parameters['b1']
    W2 = parameters['W2']
    b2 = parameters['b2']
    W3 = parameters['W3']
    b3 = parameters['b3']

    costs = []

    for i_iter in range(num_iterations):
        J = 0
        dW1 = np.zeros_like(W1)
        db1 = np.zeros_like(b1)
        dW2 = np.zeros_like(W2)
        db2 = np.zeros_like(b2)
        dW3 = np.zeros_like(W3)
        db3 = np.zeros_like(b3)

        for i in range(m):
            x = X[:, i][:, None]
            y = Y[:, i]

            # Forward propagation
            z1 = np.dot(W1, x) + b1
            a1 = relu(z1)
            z2 = np.dot(W2, a1) + b2
            a2 = relu(z2)
            z3 = np.dot(W3, a2) + b3
            a3 = sigmoid(z3)
```

(continued)

Listing 3.9 (continued)

```
            # Compute cost
            loss = compute_loss(a3, y)
            J += loss

            # Backward propagation
            dz3 = a3 - y
            dW3 += np.dot(dz3, a2.T)
            db3 += dz3

            da2 = np.dot(W3.T, dz3)
            dz2 = da2 * relu_derivative(z2)
            dW2 += np.dot(dz2, a1.T)
            db2 += dz2

            da1 = np.dot(W2.T, dz2)
            dz1 = da1 * relu_derivative(z1)
            dW1 += np.dot(dz1, x.T)
            db1 += dz1

        J /= m
        dW1 /= m
        dW2 /= m
        dW3 /= m
        db1 /= m
        db2 /= m
        db3 /= m

        # Update parameters
        W1 = W1 - learning_rate * dW1
        b1 = b1 - learning_rate * db1
        W2 = W2 - learning_rate * dW2
        b2 = b2 - learning_rate * db2
        W3 = W3 - learning_rate * dW3
        b3 = b3 - learning_rate * db3

        costs.append(np.squeeze(J))
        if i_iter % 500 == 0:
            print(f"Iteration: {i_iter}, Cost: {np.squeeze(J)}")

parameters = {"W1": W1,
        "b1": b1,
```

(continued)

Listing 3.9 (continued)
```
            "W2": W2,
            "b2": b2,
            "W3": W3,
            "b3": b3,
        }

return parameters, costs
```

The above code is a comprehensive implementation of a three-layer neural network training process. It starts by preparing the data and initializing the parameters of the network. The core of the training process is an iterative loop. In each iteration, the network undergoes forward propagation where it uses the current parameters to make predictions on the input data. The cost is calculated by comparing the predictions made by the network with the actual labels. This cost serves as a metric for evaluating the performance of the network's predictions. Following the cost computation, the network undergoes backward propagation. This is where the magic of learning happens. The network calculates the gradients of the cost with respect to the parameters. These gradients indicate how much a small change in each parameter would affect the cost.

The parameters are subsequently adjusted against the opposite direction of the gradients, which effectively minimizes the cost. This process of forward propagation, cost computation, backward propagation, and parameter update is repeated for a specified number of iterations.

Note that we are assuming a binary classification problem here for our credit classification task. Therefore, we are utilizing the binary cross-entropy loss function. This loss value is then summed and divided by the total number of examples, denoted as m, to compute the average cost for each iteration.

Throughout the training process, the cost is logged and can be visualized to monitor the learning progress. After the training is complete, the function returns the learned parameters, which can be used for making predictions on new, unseen data. This entire process embodies the essence of supervised learning in deep neural networks.

Let us now train for 2000 iterations leveraging our non-vectorized training function on `X_train` and `y_train` credit classification data prepared earlier:

```
num_iterations = 2000
learning_rate = 0.01
hidden_layer_sizes = [3, 3]
parameters_nonvec, costs = train_nonvec(X_train, y_train,
learning_rate, num_iterations, hidden_layer_sizes)
```

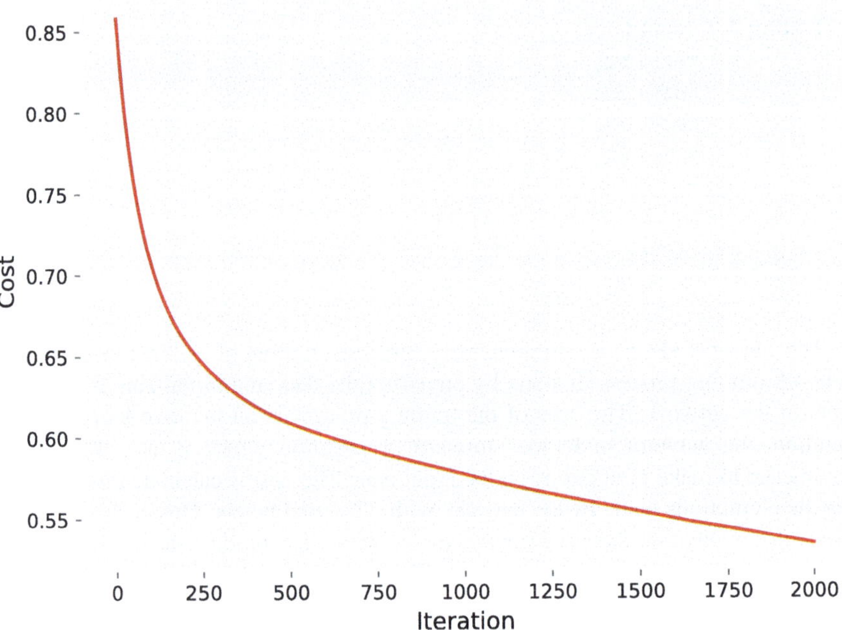

Fig. 3.6 The learning curve illustrating cost minimization. It highlights how the cost decreases over time as the model learns from the data, indicating improved performance

We can visualize the learning curve (Domhan et al. 2015), which represents the cost as a function of the number of iterations, using the following code snippet:

Listing 3.10 Plotting Costs

```
def plot_costs(costs):

    plt.figure()
    plt.plot(costs)
    plt.xlabel("Iteration")
    plt.ylabel("Cost")

    return None
```

Running the `plot_costs` function will generate the learning curve as shown in Fig. 3.6:

```
plot_costs(costs)
```

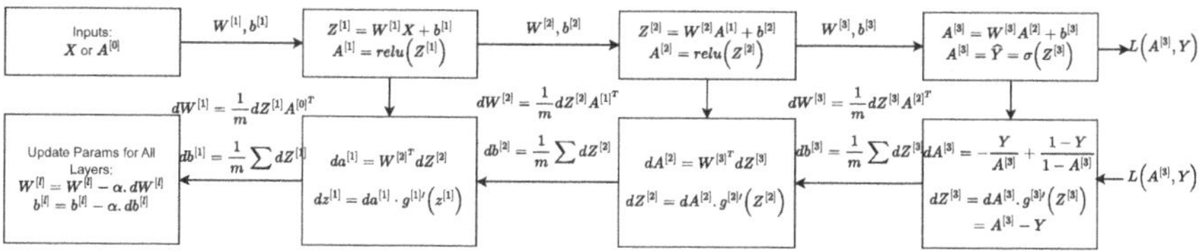

Fig. 3.7 The vectorized versions of forward and backward propagation formulas for the three-layer neural network

The learning curve effectively validates our expectation of a decreasing cost over the iterations. This trend is a positive indication, suggesting that our network is successfully learning and improving with each iteration by minimizing the cost.

Vectorized Training

The forward and backward propagation processes, as described above, require a loop to go over all training examples one by one. This implies that to train on an entire dataset, we would need to repeat these computations for each individual example and then average the cost and derivatives over all examples. While this approach is conceptually straightforward, it is computationally inefficient, especially for large datasets. This is where vectorization comes into play. Vectorization is an effective method that enables us to execute operations on whole data arrays at once, taking advantage of the parallel processing features of modern hardware. By formulating our forward and backward propagation processes in a vectorized manner, we can compute the cost and derivatives for all examples in the dataset at once, significantly improving computational efficiency.

The vectorized versions of forward and backward propagation are depicted in Fig. 3.7. By transitioning from a loop-based implementation to a vectorized one, we can greatly enhance the speed and scalability of our neural network training process. This is a crucial consideration in practical deep learning applications, where we often work with large-scale, high-dimensional data.

Next, we implement a vectorized version (Harris et al. 2020; Van Der Walt et al. 2011) of training that consists of forward and backward propagations. Let us first code up the forward propagation part in the following code:

Listing 3.11 Forward Propagation of a Neural Network

```
def forward_propagation(X, parameters):
    W1 = parameters['W1']
    b1 = parameters['b1']
    W2 = parameters['W2']
    b2 = parameters['b2']
    W3 = parameters['W3']
    b3 = parameters['b3']

    Z1 = np.dot(W1, X) + b1
    A1 = relu(Z1)
    Z2 = np.dot(W2, A1) + b2
    A2 = relu(Z2)
    Z3 = np.dot(W3, A2) + b3
    A3 = sigmoid(Z3)

    cache = (X, Z1, A1, W1, b1, Z2, A2, W2, b2, Z3, A3, W3, b3)

    return A3, cache
```

The corresponding backward propagation code will then be:

Listing 3.12 Backward Propagation of a Neural Network

```
def backward_propagation(A3, Y, cache):
    m = Y.shape[1]
    (X, Z1, A1, W1, b1, Z2, A2, W2, b2, Z3, A3, W3, b3) = cache

    dZ3 = A3 - Y
    dW3 = 1. / m * np.dot(dZ3, A2.T)
    db3 = 1. / m * np.sum(dZ3, axis=1, keepdims=True)

    dA2 = np.dot(W3.T, dZ3)
    dZ2 = dA2 * relu_derivative(Z2)
    dW2 = 1. / m * np.dot(dZ2, A1.T)
    db2 = 1. / m * np.sum(dZ2, axis=1, keepdims=True)

    dA1 = np.dot(W2.T, dZ2)
    dZ1 = dA1 * relu_derivative(Z1)
    dW1 = 1. / m * np.dot(dZ1, X.T)
    db1 = 1. / m * np.sum(dZ1, axis=1, keepdims=True)
```

(continued)

Listing 3.12 (continued)

```
    grads = {"dZ3": dZ3, "dW3": dW3, "db3": db3,
             "dA2": dA2, "dZ2": dZ2, "dW2": dW2, "db2": db2,
             "dA1": dA1, "dZ1": dZ1, "dW1": dW1, "db1": db1}

    return grads
```

The parameters will be updated as follows:

Listing 3.13 Updating Parameters in Neural Network Training

```
def update_parameters(parameters, grads, learning_rate):

    W1 = parameters['W1']
    b1 = parameters['b1']
    W2 = parameters['W2']
    b2 = parameters['b2']
    W3 = parameters['W3']
    b3 = parameters['b3']

    dW1 = grads['dW1']
    db1 = grads['db1']
    dW2 = grads['dW2']
    db2 = grads['db2']
    dW3 = grads['dW3']
    db3 = grads['db3']

    W1 = W1 - learning_rate * dW1
    b1 = b1 - learning_rate * db1
    W2 = W2 - learning_rate * dW2
    b2 = b2 - learning_rate * db2
    W3 = W3 - learning_rate * dW3
    b3 = b3 - learning_rate * db3

    parameters = {"W1": W1,
                  "b1": b1,
                  "W2": W2,
                  "b2": b2,
                  "W3": W3,
                  "b3": b3,
                 }

    return parameters
```

Finally, we put all the training components together in the following code:

Listing 3.14 Non-vectorized Training of a Neural Network Model

```
def train_vec(X, y, learning_rate, num_iterations, hidden_layer_
sizes):

    X = X.T
    Y = y.reshape(1, -1)
    nx, m = X.shape

    layer_sizes = [nx] + hidden_layer_sizes + [1]
    parameters = initialize_parameters(layer_sizes)

    costs = []

    for i_iter in range(num_iterations):
        A3, cache = forward_propagation(X, parameters)

        cost = compute_cost(A3, Y)

        grads = backward_propagation(A3, Y, cache)

        parameters = update_parameters(parameters, grads,
learning_rate)

        costs.append(np.squeeze(cost))
        if i_iter % 500 == 0:
            print(f"Iteration: {i_iter}, Cost: {np.squeeze(cost)}")

    return parameters, costs
```

As we have previously discussed, non-vectorized training lacks efficiency. This inefficiency stems from the fact that we process one example at a time. In contrast, the vectorized version enhances efficiency by performing forward propagation, gradient calculation, and parameter updates in a single pass for all examples. This approach significantly accelerates the learning process.

Similarly, we now train for 2000 iterations using the same neural network architecture:

```
num_iterations = 2000
learning_rate = 0.01
hidden_layer_sizes = [3, 3]
parameters_vec, costs = train_vec(X_train, y_train, learning_rate,
num_iterations, hidden_layer_sizes)
```

3.5.6 Prediction

The final step in a neural network is prediction. Prediction is nothing but one pass of forward propagation with the trained weights and biases. This process involves applying the weights and biases to the input data and passing the result through the activation functions of each layer. The final layer's output is the network's prediction. The network's performance can be evaluated by comparing its predictions with the actual target values. A well-trained network should demonstrate its ability to generalize from the training data to new, unknown situations by making accurate predictions on unseen data. This is the ultimate goal of training a neural network.

Non-vectorized Prediction

The following code shows how to perform non-vectorized prediction using the trained parameters:

Listing 3.15 Non-vectorized Prediction of a Neural Network Model

```
def predict_nonvec(X, parameters):

    W1 = parameters['W1']
    b1 = parameters['b1']
    W2 = parameters['W2']
    b2 = parameters['b2']
    W3 = parameters['W3']
    b3 = parameters['b3']

    X = X.T
    nx, m = X.shape

    y_hat = []
    for i in range(m):
        x = X[:, i][:, None]

        z1 = np.dot(W1, x) + b1
        a1 = relu(z1)
        z2 = np.dot(W2, a1) + b2
        a2 = relu(z2)
        z3 = np.dot(W3, a2) + b3
        a3 = sigmoid(z3)

        y_hat.append(float(np.squeeze(a3)))

    return np.array(y_hat)
```

It is time for prediction:

```
y_hat = predict_nonvec(X_test, parameters_nonvec)
y_pred = (y_hat>0.5).astype(int)
```

Finally, for the sake of validation, we can produce classification metrics using Sklearn:

Listing 3.16 Producing Classification Metrics

```
from sklearn.metrics import f1_score, precision_score, recall_
score

def classification_metrics(y_test, y_pred):
    f1 = f1_score(y_test, y_pred)
    precision = precision_score(y_test, y_pred)
    recall = recall_score(y_test, y_pred)

    print(f'F1 Score: {f1}')
    print(f'Precision: {precision}')
    print(f'Recall: {recall}')

    return precision, recall, f1
```

The following output confirms that we have achieved an F1 score (Pedregosa et al. 2011; Van Rijsbergen 1979) of 0.83 on our test (validation) set:

```
precision, recall, f1 = classification_metrics(y_test, y_pred)
```

```
F1 Score: 0.83125
Precision: 0.7388888888888889
Recall: 0.95
```

Vectorized Prediction

Similarly, here is the code that implements the prediction in vectorized format:

Listing 3.17 Vectorized Prediction of a Neural Network Model

```
def predict_vec(X, parameters):

    X = X.T
    y_hat, _ = forward_propagation(X, parameters)
    y_hat = np.squeeze(y_hat)

    return y_hat
```

In a similar way, we execute vectorized predictions using the following code snippet and generate classification metrics utilizing the parameters obtained from the training phase:

```
y_hat = predict_vec(X_test, parameters_vec)
y_pred = (y_hat>0.5).astype(int)
precision, recall, f1 = classification_metrics(y_test, y_pred)
```

```
F1 Score: 0.83125
Precision: 0.7388888888888889
Recall: 0.95
```

We can see that the F1 score is identical with our non-vectorized implementation. This is what we expect. This means, we have successfully developed an efficient training process for building a deep neural network from start to finish. We have successfully developed a credit risk classification algorithm. Now, we can predict the credit risk for any incoming unseen data. This will enable us to classify individuals into categories of good or bad credit risks.

3.6 Summary

- Neural network training involves key components such as data, layers, loss function, and optimization algorithm, which collectively enable the network to learn complex patterns from data.
- The architecture of a neural network comprises input, hidden, and output layers, with each layer playing a specific role in processing information and making predictions.
- Activation functions introduce nonlinearity into neural networks, crucial for learning complex relationships in data, with popular functions including Sigmoid, tanh, ReLU, and Leaky ReLU.

- Matrix dimensions and parameters are essential considerations in neural network design, dictating the structure and size of weight matrices and bias vectors across layers.
- Training a neural network follows steps such as normalization of input features, initialization of parameters, defining a cost function, iterative optimization through forward and backward propagation, and making predictions using trained parameters.
- A practical example of credit classification demonstrates the application of neural network training techniques, including feature normalization, non-vectorized and vectorized training processes, and visualization of learning curves for monitoring model performance.

References

Bengio, Y., Simard, P., & Frasconi, P. (1994). Learning long-term dependencies with gradient descent is difficult. *IEEE Transactions on Neural Networks*, *5*(2), 157–166.

Domhan, T., Springenberg, J. T., & Hutter, F. (2015). Speeding up automatic hyperparameter optimization of deep neural networks by extrapolation of learning curves. *IJCAI*, *15*, 3460–3468.

Glorot, X., & Bengio, Y. (2010). Understanding the difficulty of training deep feedforward neural networks. *Proceedings of the Thirteenth International Conference on Artificial Intelligence and Statistics*, 249–256.

Goodfellow, I. (2016). *Deep learning*. MIT press.

Halevy, A., Norvig, P., & Pereira, F. (2009). The unreasonable effectiveness of data. *IEEE Intelligent Systems*, *24*(2), 8–12.

Hand, D. J., & Henley, W. E. (1997). Statistical classification methods in consumer credit scoring: A review. *Journal of the Royal Statistical Society: Series a (Statistics in Society)*, *160*(3), 523–541.

Harris, C. R., Millman, K. J., Van Der Walt, S. J., Gommers, R., Virtanen, P., Cournapeau, D., Wieser, E., Taylor, J., Berg, S., Smith, N. J., & others. (2020). Array programming with NumPy. *Nature*, *585*(7825), 357–362.

He, K., Zhang, X., Ren, S., & Sun, J. (2015). Delving deep into rectifiers: Surpassing human-level performance on ImageNet classification. *Proceedings of the IEEE International Conference on Computer Vision*, 1026–1034.

LeCun, Y., Bengio, Y., & Hinton, G. (2015). Deep learning. *Nature*, *521*(7553), 436–444.

LeCun, Y., Bottou, L., Orr, G. B., & Müller, K.-R. (2002). Efficient backprop. In *Neural networks: Tricks of the trade* (pp. 9–50). Springer.

Maas, A. L., Hannun, A. Y., Ng, A. Y., & others. (2013). Rectifier nonlinearities improve neural network acoustic models. *Proc. ICML*, *30*(1), 3.

Nair, V., & Hinton, G. E. (2010). Rectified linear units improve restricted Boltzmann machines. *Proceedings of the 27th International Conference on Machine Learning (ICML-10)*, 807–814.

Pedregosa, F., Varoquaux, G., Gramfort, A., Michel, V., Thirion, B., Grisel, O., Blondel, M., Prettenhofer, P., Weiss, R., Dubourg, V., & others. (2011). Scikit-learn: Machine learning in Python. *The Journal of Machine Learning Research*, *12*, 2825–2830.

Ruder, S. (2016). An overview of gradient descent optimization algorithms. *arXiv Preprint arXiv:1609.04747*.

Schmidhuber, J. (2015). Deep learning in neural networks: An overview. *Neural Networks*, *61*, 85–117.

Spivak, M. (1994). Calculus, Publish or Perish. *Inc., Houston, Texas.*

Sutskever, I., Martens, J., Dahl, G., & Hinton, G. (2013). On the importance of initialization and momentum in deep learning. *International Conference on Machine Learning*, 1139–1147.

Van Der Walt, S., Colbert, S. C., & Varoquaux, G. (2011). The NumPy array: A structure for efficient numerical computation. *Computing in Science & Engineering*, *13*(2), 22–30.

Van Rijsbergen, C. J. (1979). *Information Retrieval, 2nd edn. Newton, MA*. USA: Butterworth-Heinemann.

Werbos, P. J. (1994). *The roots of backpropagation: From ordered derivatives to neural networks and political forecasting*. John Wiley & Sons.

Chapter 4
Dealing with Bias and Variance

This chapter covers

- Understanding the concepts of bias and variance.
- Exploring L2 regularization technique to control model complexity.
- Implementing dropout regularization to prevent overfitting.
- Incorporating early stopping in training.
- Demonstrating the practical application of regularization and dropout.

In this chapter, we dive into the critical aspects of bias and variance that influence the performance of deep learning models. Understanding these concepts is essential for effectively managing model complexity and preventing overfitting, which can compromise generalization—the model's ability to perform well on unseen data. We will explore various regularization techniques, including L2 regularization, dropout, and early stopping, to strike a balance between fitting the training data and generalizing to unseen data. Through practical implementations and examples, we will demonstrate how regularization methods can enhance model robustness and reduce model complexity.

4.1 Bias and Variance

The concept of the bias–variance trade-off (Geman et al. 1992) is key in the development of efficient deep learning models. Bias is the error that arises when we simplify a complex real-world problem by using a less complicated model. For example, if the real-world problem is nonlinear, approximating it with a basic linear model would lead to high bias, a situation commonly referred to as under-fitting (Hastie et al. 2009). Conversely, variance measures an algorithm's sensitivity to variations in the training data. A model with high variance learns not only the underlying pattern but also the noise present in the training data, a phenomenon known as overfitting (Hawkins 2004). The ultimate goal in building a deep learning

T. Islam, *Hands-on Deep Learning*, https://doi.org/10.1007/978-3-032-00488-8_4

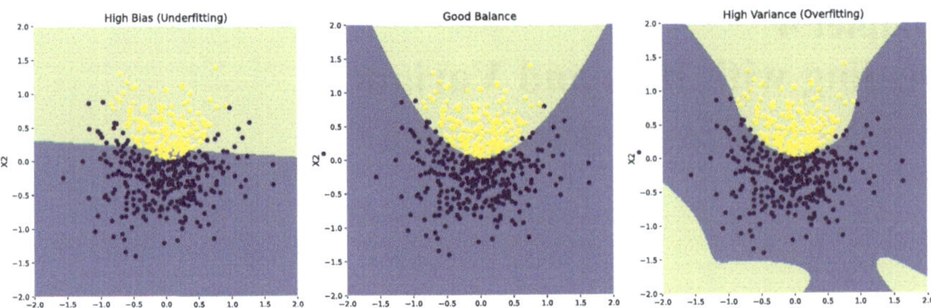

Fig. 4.1 The concept of the bias–variance trade-off. On the left, high bias indicates an overly simplistic model. On the right, high variance signifies an overly complex model. The ideal model complexity is in the middle, striking a balance between bias and variance

model is to strike a balance between high bias (underfitting) and high variance (overfitting) (Domingos 2012). This ensures that the model performs optimally on both the training data (seen data) and any new data it encounters (unseen data). This delicate balancing act is what we refer to as the bias–variance trade-off.

The concept of the bias–variance trade-off is illustrated in Fig. 4.1, which depicts a binary classification scenario. The linear decision boundary of the high-bias model exemplifies underfitting, as the actual decision boundary in this case is nonlinear. In comparison, the decision boundary of the high-variance model is overly complex, as it attempts to fit every point in the training set, including noise. A more balanced approach is represented by a smoother, nonlinear decision boundary, as shown in the middle panel. This decision boundary strikes a good balance between bias and variance, avoiding both underfitting and overfitting.

In building deep neural networks, bias and variance are the two types of errors that will guide us to determine if the model is trained correctly or not. The bias and variance errors are generally determined by employing a training and a validation set. There is another set called test set that is used to verify "generalization" of the model.

Here is a rule-of-thumb guideline for the proportion of training, validation, and test sets to be used while training a deep learning model. A common approach for splitting data into these sets is a 70:15:15 split when the data size is small. In the case of a large data size, the proportions for the validation and test sets can be significantly reduced, thus a 95:3:2 split is also acceptable.

Generally speaking,

- Train set will provide information on "bias."
- Validation set will provide information on "variance."
- Test set will provide information on the "generalization" of the model.

Let us assume few scenarios where we have train and validation set errors for ImageNet image classification (Table 4.1). We also know that human level performance on this dataset is around 5%. In the table, model 1 suffers high bias problem. In model 2, the error rate in training set is reasonable; however, error rate in

Table 4.1 Examples illustrating bias/variance scenarios

Model no.	Train set error (%)	Validation set error (%)	Bias/variance
1	25	28	High bias
2	2	20	High variance
3	25	50	High bias and high variance
4	4	5	Low bias and low variance

validation set is rather high. It suffers high variance problem. The model 3's training set error is high and validation set error is 2x of training set error. Therefore, it suffers both high bias and high variance problems. Model 4 seems to strike a good balance with low bias and low variance.

To effectively address bias–variance dilemmas, it is advisable that we adopt a systematic, step-by-step dissection approach. The initial step in this process involves the identification of a bias problem, which is primarily determined by examining the performance of the training set. After addressing the bias problem, the next step is to identify the variance problem by checking the validation set's performance.

If the model's performance on the training set is subpar, it is indicative of a high bias issue. To mitigate this, there are several potential strategies that can be employed:

- *Increasing model complexity*: This involves increasing the number of parameters in the network, thereby making model or architecture more complex.
- *Extending training duration*: If the loss is observed to be decreasing, yet the training set error remains high, prolonging the training duration could potentially enhance the model's performance.

When the model's performance on the training set is satisfactory, but its performance on the validation set is less than ideal, this typically signals a high variance issue. To address this, there are several potential strategies that can be implemented:

- *Reducing model complexity*: This strategy involves decreasing the number of parameters in the network, thereby simplifying the model or architecture. This can be achieved through methods such as architecture tuning, regularization, or the incorporation of dropout layers.
- *Increasing training data*: Another effective approach is to train the model with additional data. This can help the model generalize better, thereby improving its performance on the validation set.

4.2 Regularization

As we discussed in the previous section, one effective strategy to combat variance or overfitting is to reduce model complexity. Regularization is a technique that serves this purpose (Krogh and Hertz 1991). The essence of regularization lies in shrinking the weights toward zero (Hoerl and Kennard 1970). Consider a single node in a neural network with five input features. The computation of the weighted sum vector, denoted as z, is as follows:

$$z^{(i)} = w^T x^{(i)} + b$$

$$z = w_1 x_1 + w_2 x_2 + w_3 x_3 + w_4 x_4 + w_5 x_5 + b$$

Now, if we can somehow shrink (eliminate) the "w" parameters for some of the features, say for "w_2" and "w_3," the above computation simplifies to:

$$z = w_1 x_1 + w_4 x_4 + w_5 x_5 + b$$

This demonstrates how regularization can effectively simplify the model by reducing the influence of certain features. This, in turn, helps to mitigate overfitting and improve the model's generalization capability.

The next question then is, how can we achieve this? Interestingly, this can be accomplished by regularizing the cost function. We will demonstrate this through a real-world example.

4.2.1 KDD Network Intrusion Classification

In this example, we will be using KDD Cup 1999 network intrusion dataset (Tavallaee et al. 2009). Specifically, we will be using the "SA" subset. To make our analysis manageable and efficient, we are working with 10% of the data. The labels in the dataset represent different types of network interactions, categorized as either "normal" or various types of "attacks." We have formulated this as a binary classification problem, where "y" is our target variable. In this context, "y" represents whether a network connection is "normal" or not. We label "normal" connections as 0 and all other types of connections, which are potential intrusions, as 1. This simplifies our problem to a "normal" versus "others" classification, enabling us to focus on distinguishing between regular network behavior and potential threats.

Here is the code that loads the kddcup99 data:

Listing 4.1 Loading KDD Cup 1999 Network Intrusion Data

```
from sklearn.datasets import fetch_kddcup99

def load_kddcup_data():

    X, y = fetch_kddcup99(
        subset="SA", percent10=True, random_state=42,
```

(continued)

Listing 4.1 (continued)

```
return_X_y=True, as_frame=True
    )
    y = (y != b"normal.").astype(np.int32).values

    features = ['src_bytes',
               'dst_bytes', 'land', 'wrong_fragment', 'urgent', 'hot',
               'num_failed_logins', 'logged_in', 'num_compromised',
'root_shell',
               'su_attempted', 'num_root', 'num_file_creations',
'num_shells',
               'num_access_files', 'num_outbound_cmds', 'is_host_login',
               'is_guest_login', 'count', 'srv_count', 'serror_rate',
               'srv_serror_rate', 'rerror_rate', 'srv_rerror_rate',
'same_srv_rate',
               'diff_srv_rate', 'srv_diff_host_rate', 'dst_host_count',
               'dst_host_srv_count', 'dst_host_same_srv_rate',
               'dst_host_diff_srv_rate', 'dst_host_same_src_port_rate',
               'dst_host_srv_diff_host_rate', 'dst_host_serror_rate',
               'dst_host_srv_serror_rate', 'dst_host_rerror_rate',
               'dst_host_srv_rerror_rate']
    X = X[features].astype(float)

    return X, y
```

Let us now load the data:

```
X, y = load_kddcup_data()
```

Next, we partition the dataset into training and testing subsets, and normalize the input features:

```
X_train, X_test, y_train, y_test = split_dataset(X, y)
X_train, X_test = normalize_features(X_train, X_test)
```

4.2.2 Regularizing the Cost Function

As we learned from the previous chapter, the cross-entropy cost function for binary classification is defined as:

$$J_{\text{cross entropy cost}} = -\frac{1}{m}\sum_{i=1}^{m}\left[y^{(i)}\log\left(a^{[L](i)}\right) + \left(1 - y^{(i)}\right)\log\left(1 - a^{[L](i)}\right)\right]$$

Now, we introduce a new L2 regularization cost function (Ng 2004) to be added to the above cost function as:

$$J_{\text{L2 regularization cost}} = \frac{1}{m}\frac{\lambda}{2}\sum_{l}\sum_{k}\sum_{j}W_{k,j}^{[l]2}$$

Therefore, final regularized cost function becomes:

$$J_{regularized} = J_{\text{cross entropy cost}} + J_{\text{L2 regularization cost}}$$

We can generalize this for any given loss function:

$$J(W, b) = \frac{1}{m}\sum_{i=1}^{m}\mathcal{L}\left(y^{(i)}, \widehat{y}^{(i)}\right) + \frac{\lambda}{2m}\sum_{l=1}^{L}||W^{(l)}||_F^2$$

where:

- $\mathcal{L}\left(y^{(i)}, \widehat{y}^{(i)}\right)$ is the loss function for the ith example,
- λ is the regularization parameter,
- m is the number of examples,
- L is the number of layers in the network,
- $||W^{(l)}||_F$ is the Frobenius norm of the weight matrix (Golub and Van Loan 2013) in the l th layer,
- The sum is over all layers in the network.

From the above equation, we can see that only W parameters are regularized. It is indeed possible to apply regularization to the bias terms b as well. However, these terms are typically of a lower dimension compared to the weight matrices (W), and their contribution to overfitting is generally considered to be minimal. Therefore, in practice, we often ignore the regularization of the bias terms.

The regularization parameter, λ is a hyperparameter that is responsible for controlling the complexity of the model. A higher value of lambda will impose a greater penalty on larger weights, effectively shrinking them toward zero. By driving weights toward zero, the model's complexity is reduced, which thereby helps to prevent overfitting.

Let us now implement the modified cost function including the regularization term in the following code using the above equations:

Listing 4.2 Computing Cost with L2 Regularization

```
def compute_cost(Y_hat, Y, parameters, lambda_):

    W1 = parameters['W1']
    b1 = parameters['b1']
    W2 = parameters['W2']
    b2 = parameters['b2']
    W3 = parameters['W3']
    b3 = parameters['b3']

    m = Y.shape[1]

    epsilon = 1e-10
    cross_entropy_cost = -(1/m) * np.sum(Y * np.log(Y_hat +
epsilon) + (1 - Y) * np.log(1 - Y_hat + epsilon))

   regularization_cost = lambda_/(2*m) * (np.sum(W1**2) + np.sum
(W2**2) + np.sum(W3**2))

    cost = cross_entropy_cost + regularization_cost

    return cost
```

Now that we have modified our cost function for L2 regularization, let us revisit the forward propagation for the three-layer fully connected neural network, which remains unchanged from the previous chapter:

Listing 4.3 Forward Propagation with L2 Regularization

```
def forward_propagation(X, parameters):
    W1 = parameters['W1']
    b1 = parameters['b1']
    W2 = parameters['W2']
    b2 = parameters['b2']
    W3 = parameters['W3']
    b3 = parameters['b3']

    Z1 = np.dot(W1, X) + b1
    A1 = relu(Z1)
    Z2 = np.dot(W2, A1) + b2
    A2 = relu(Z2)
```

(continued)

Listing 4.3 (continued)

```
        Z3 = np.dot(W3, A2) + b3
        A3 = sigmoid(Z3)

        cache = (X, Z1, A1, W1, b1, Z2, A2, W2, b2, Z3, A3, W3, b3)

        return A3, cache
```

However, the backpropagation that is partial derivative of cost function would change since we have a modified cost function. From the previous chapter, we know:

$$dW^{[l]} = \frac{1}{m} dZ^{[l]} A^{[l-1]T}$$

For L2 regularization, the above equation can be expanded as:

$$dW^{[l]} = \frac{1}{m} dZ^{[l]} A^{[l-1]T} + \frac{\lambda}{m} W^{[l]}$$

Then,

$$W^{[l]} = W^{[l]} - \alpha.dW$$

The code snippet provided below represents a custom implementation of backward propagation with L2 regularization. It is important to note that since we have incorporated a regularization term into the cost function, the partial derivative of the cost with respect to the weights (dW1, dW2, and dW3) will be adjusted accordingly:

Listing 4.4 Backward Propagation with L2 Regularization

```
def backward_propagation(A3, Y, lambda_, cache):
        m = Y.shape[1]
        (X, Z1, A1, W1, b1, Z2, A2, W2, b2, Z3, A3, W3, b3) = cache

        dZ3 = A3 - Y
        dW3 = 1. / m * np.dot(dZ3, A2.T) + lambda_/m * W3
        db3 = 1. / m * np.sum(dZ3, axis=1, keepdims=True)

        dA2 = np.dot(W3.T, dZ3)
        dZ2 = dA2 * relu_derivative(Z2)
```

(continued)

Listing 4.4 (continued)
```
    dW2 = 1. / m * np.dot(dZ2, A1.T) + lambda_/m * W2
    db2 = 1. / m * np.sum(dZ2, axis=1, keepdims=True)

    dA1 = np.dot(W2.T, dZ2)
    dZ1 = dA1 * relu_derivative(Z1)
    dW1 = 1. / m * np.dot(dZ1, X.T) + lambda_/m * W1
    db1 = 1. / m * np.sum(dZ1, axis=1, keepdims=True)

    grads = {"dZ3": dZ3, "dW3": dW3, "db3": db3,
             "dA2": dA2, "dZ2": dZ2, "dW2": dW2, "db2": db2,
             "dA1": dA1, "dZ1": dZ1, "dW1": dW1, "db1": db1}

    return grads
```

4.2.3 Training

Let us now work on implementing the training function with regularization. The below code defines a training function for a neural network, incorporating regularization to prevent overfitting. The `lambda_` parameter is used in the `compute_cost` and `backward_propagation` functions to apply regularization, which penalizes large weights and helps the model generalize better. By including `lambda_` in the cost calculation and gradient updates, the model aims to balance fitting the training data and maintaining simplicity, ultimately improving performance on unseen data:

Listing 4.5 Training Neural Network with L2 Regularization
```
def train(X, y, learning_rate, num_iterations, hidden_layer_
sizes, lambda_):

    X = X.T
    Y = y.reshape(1, -1)
    nx, m = X.shape

    layer_sizes = [nx] + hidden_layer_sizes + [1]
    print (layer_sizes)
    parameters = initialize_parameters(layer_sizes)

    costs = []
```
(continued)

Listing 4.5 (continued)

```
    for i_iter in range(num_iterations):
        A3, cache = forward_propagation(X, parameters)

        cost = compute_cost(A3, Y, parameters, lambda_)

        grads = backward_propagation(A3, Y, lambda_, cache)

        parameters = update_parameters(parameters, grads,
learning_rate)

        costs.append(np.squeeze(cost))
        if i_iter % 100 == 0:
            print(f"Iteration: {i_iter}, Cost: {np.squeeze(cost)}")

    return parameters, costs
```

Now, let us train a three-layer neural network without any regularization. Setting `lambda_` to zero effectively means that no regularization is being applied:

```
num_iterations = 1000
learning_rate = 0.01
hidden_layer_sizes = [3, 3]
lambda_ = 0.0
parameters, costs = train(X_train, y_train, learning_rate,
num_iterations, hidden_layer_sizes, lambda_)
```

Let us inspect the parameters trained. The following code snippet allows us to compute and display the average value of these parameters:

```
print ("Mean W1 = ", parameters['W1'].mean())
print ("Mean W2 = ", parameters['W2'].mean())
print ("Mean W3 = ", parameters['W2'].mean())
```

```
Mean W1 = -0.03202634966979897
Mean W2 = -0.19708551876039437
Mean W3 = -0.19708551876039437
```

Next, we train the same network, but this time with a regularization:

```
lambda_ = 100
parameters, costs = train(X_train, y_train, learning_rate,
num_iterations, hidden_layer_sizes, lambda_)
print ("Mean W1 = ", parameters['W1'].mean())
print ("Mean W2 = ", parameters['W2'].mean())
print ("Mean W3 = ", parameters['W2'].mean())
```

This time, it generates the following outputs:

```
Mean W1 = -0.03154972729335642
Mean W2 = -0.19407093812660262
Mean W3 = -0.19407093812660262
```

The difference in the mean values of the parameters between the models with no regularization and L2 regularization is due to the effect of the regularization term. L2 regularization is expected to produce smaller weights, which is why the mean of the parameters is slightly smaller when using L2 regularization. In our case, the mean of the weights W1, W2, and W3 in the model with L2 regularization is closer to zero compared to the model with no regularization. This is consistent with the effect of L2 regularization, which tends to make the weights shrink toward zero.

4.2.4 Prediction and Validation

When we apply regularization, the prediction remains unaffected. We can utilize the same "`predict`" and "`classification_metrics`" functions that we implemented in the previous chapter:

```
y_hat = predict(X_test, parameters)
y_pred = (y_hat>0.5).astype(int)
precision, recall, f1 = classification_metrics(y_test, y_pred)
```

```
F1 Score: 0.9626239511823036
Precision: 0.9546142208774584
Recall: 0.9707692307692307
```

From the generated output, we observe that the F1-score (Pedregosa et al. 2011; Van Rijsbergen 1979) on the test (validation) set is an impressive 0.96. This means we have successfully built a neural network classifier from scratch incorporating regularization. Our model is capable of differentiating between standard network behavior and potential threats.

4.3 Dropout

Dropout (Srivastava et al. 2014) is another regularization technique used in neural networks to prevent overfitting. The dropout technique randomly "drops out" (i.e., sets to zero) a number of activation nodes in the hidden layers during training phase (Baldi and Sadowski 2013). How many nodes or units will be "dropped out" is determined by dropout rate, which is a hyperparameter. Practically, this is done by multiplying the output value of each neuron by a random variable that takes the mask value 0 (with probability equal to the dropout rate) or 1 (with probability equal to 1—dropout rate). An illustrative figure of dropout is shown in Fig. 4.2. In this figure, the activation nodes in red color are dropped out for a single pass. This means in this pass, these activation nodes will not be used in the training.

From previous chapter, we know that non-vectorized version of forward propagation for a single example is as follows:

$$z^{[l]} = W^{[l]}a^{[l-1]} + b^{[l]}$$

$$a^{[l]} = g^{[l]}\left(z^{[l]}\right)$$

When dropout is applied during forward propagation, we first create a dropout mask. This is done by creating random samples from a uniform distribution over 0 and 1:

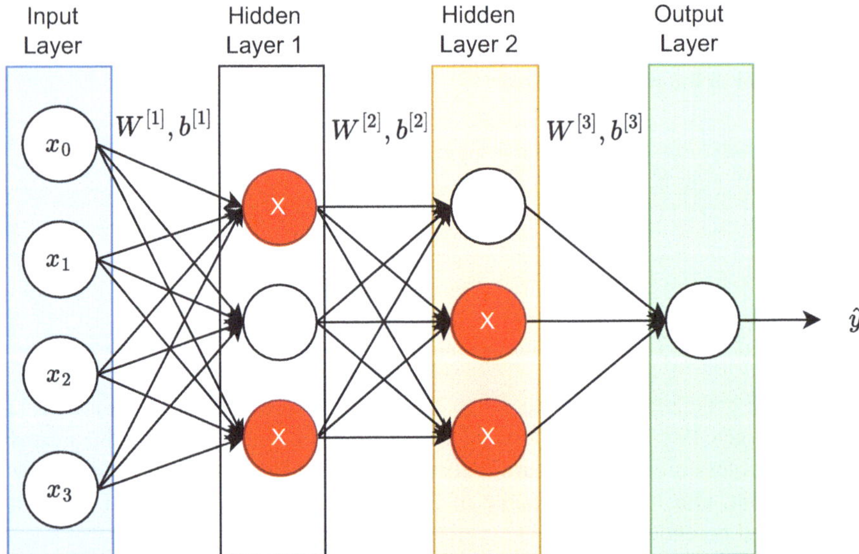

Fig. 4.2 An illustrative figure of dropout. This figure demonstrates the concept of dropout in neural networks, where randomly selected neurons are ignored (dropped out) during training

$$d^{[l]} = random.\text{rand}(n[l], m)) \geq dropoutRate$$

Then, activation matrix is updated as follows:

$$a^{[l]} = a^{[l]} * d^{[l]}$$

However, since we are dropping out few neurons in training, this would mean that during prediction, the network would see more active neurons than it did during training. To compensate for this, we scale up the outputs of the remaining neurons during training by the same factor we used for dropout. This ensures that during testing, when all neurons are active, the total output is approximately the same as the expected output during training. It is a way of making sure the network's behavior during training is as close as possible to its behavior during testing, despite the dropout. This technique is known as "inverted dropout." This is why we scale the activation value:

$$a^{[l]} = \frac{a^{[l]}}{(1 - dropoutRate)}$$

The Python implementation will look like this:

```
a = np.random.randn(10, 1)
dropout_rate = 0.5
d = (np.random.rand(*a.shape) >= dropout_rate).astype(float)
a *= d
a /= (1 - dropout_rate)
```

4.3.1 Horses or Humans Classification

We know the drill. Now, it is time to implement dropout and apply it to a real-world use case. In this example, we will be using horses or humans data from TensorFlow dataset. This is an interesting image classification problem. The collection comprises 500 computer-generated illustrations featuring diverse breeds of horses in a variety of poses and settings. Additionally, it encompasses 527 computer-rendered depictions of people in assorted postures and environments. Figure 4.3 illustrates some example images from the dataset. Although it is an image classification problem, we can solve this using fully connected neural networks.

Let us first create a function to load data from TensorFlow dataset:

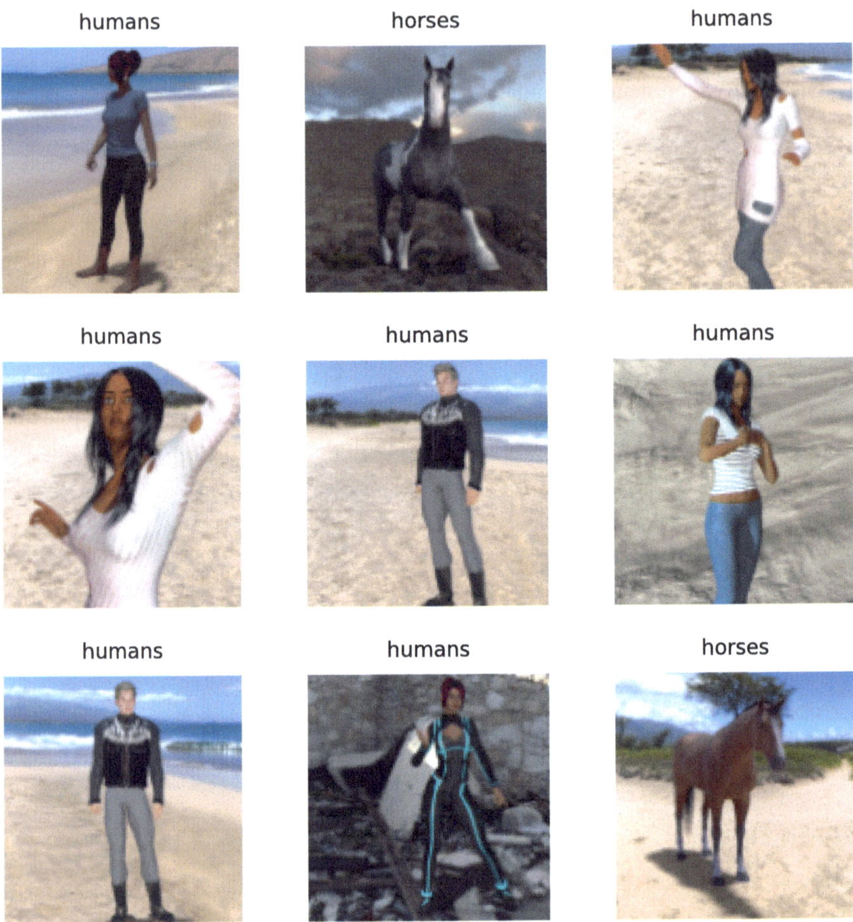

Fig. 4.3 Example images from the horses or humans dataset

Listing 4.6 Loading Horses or Humans Data

```
import tensorflow_datasets as tfds
import tensorflow as tf

def load_horses_humans_data():

    dataset, dataset_info = tfds.load('horses_or_humans',
with_info=True, split='train')
```

(continued)

Listing 4.6 (continued)
```
    X = []
    y = []

    for example in dataset:
        X.append(example['image'])
        y.append(example['label'])

    X = tf.image.resize(X, [28,28])

    X = tf.image.rgb_to_grayscale(X)

    X = np.array(X)
    y = np.array(y)
    X = X.reshape((X.shape[0], np.prod(X.shape[1:])))

    return X, y
```

We can then call the function to create X and y:

```
X, y = load_horses_humans_data()
```

We will then normalize the input features and partition the data into train and test set:

```
X = X/255.0
X_train, X_test, y_train, y_test = split_dataset(X, y)
```

4.3.2 Forward and Backward Propagations with Dropout

Let us now implement forward propagation with dropout for a three-layer neural network:

Listing 4.7 Forward Propagation with Dropout
```
def forward_propagation(X, parameters, dropout_rate):

    W1 = parameters['W1']
    b1 = parameters['b1']
```

(continued)

Listing 4.7 (continued)

```
    W2 = parameters['W2']
    b2 = parameters['b2']
    W3 = parameters['W3']
    b3 = parameters['b3']

    Z1 = np.dot(W1, X) + b1
    A1 = relu(Z1)
    D1 = (np.random.rand(*A1.shape) >= dropout_rate).astype(float)
    A1 = A1 * D1
    A1 = A1 / (1 - dropout_rate)

    Z2 = np.dot(W2, A1) + b2
    A2 = relu(Z2)
    D2 = (np.random.rand(*A2.shape) >= dropout_rate).astype(float)
    A2 = A2 * D2
    A2 = A2 / (1 - dropout_rate)

    Z3 = np.dot(W3, A2) + b3
    A3 = sigmoid(Z3)

    cache = (X, Z1, D1, A1, W1, b1, Z2, D2, A2, W2, b2, Z3, A3, W3, b3,
dropout_rate)

    return A3, cache
```

Similarly, during backpropagation, we propagate the gradients back through the network to update the parameters. Normally, the computation for the gradients for a single example would be as follows:

$$dz^{[l]} = da^{[l]} * g'^{[l]}\left(z^{[l]}\right)$$

$$dW^{[l]} = \frac{1}{m} dz^{[l]} a^{[l-1]T}$$

$$db^{[l]} = dz^{[l]}$$

$$da^{[l-1]} = W^{[l]T} dz^{[l]}$$

However, when dropout is applied, we need to account for the dropout mask $d^{[l]}$ that was applied during forward propagation. We do this by simply applying the same mask to $da^{[l]}$ during backpropagation:

$$da^{[l]} = da^{[l]} * d^{[l]}$$

And then, we scale the value of $da^{[l]}$ by the same factor as during forward propagation:

$$da^{[l]} = \frac{da^{[l]}}{(1 - dropoutRate)}$$

This ensures that the expected value of $a^{[l]}$ remains in the range as it was in forward propagation, preserving the integrity of the backpropagation process. The rest of the backpropagation process remains the same.

The Python implementation will look like this:

```
da = np.random.randn(10, 1)
da = da * d
da = da/(1 - dropout_rate)
```

Next, we implement backpropagation for the three-layer neural network:

Listing 4.8 Backward Propagation with Dropout

```
def backward_propagation(A3, Y, lambda_, cache):
    m = Y.shape[1]
    (X, Z1, D1, A1, W1, b1, Z2, D2, A2, W2, b2, Z3, A3, W3, b3,
dropout_rate) = cache

    dZ3 = A3 - Y
    dW3 = 1. / m * np.dot(dZ3, A2.T) + lambda_/m * W3
    db3 = 1. / m * np.sum(dZ3, axis=1, keepdims=True)

    dA2 = np.dot(W3.T, dZ3)
    dA2 = dA2 * D2
    dA2 = dA2/(1 - dropout_rate)
    dZ2 = dA2 * relu_derivative(Z2)
    dW2 = 1. / m * np.dot(dZ2, A1.T) + lambda_/m * W2
    db2 = 1. / m * np.sum(dZ2, axis=1, keepdims=True)

    dA1 = np.dot(W2.T, dZ2)
    dA1 = dA1 * D1
    dA1 = dA1/(1 - dropout_rate)
    dZ1 = dA1 * relu_derivative(Z1)
    dW1 = 1. / m * np.dot(dZ1, X.T) + lambda_/m * W1
    db1 = 1. / m * np.sum(dZ1, axis=1, keepdims=True)
```

(continued)

Listing 4.8 (continued)

```
    grads = {"dZ3": dZ3, "dW3": dW3, "db3": db3,
             "dA2": dA2, "dZ2": dZ2, "dW2": dW2, "db2": db2,
             "dA1": dA1, "dZ1": dZ1, "dW1": dW1, "db1": db1}

    return grads
```

4.3.3 *Training*

We now proceed to train the neural network on horse and human dataset. We will add dropout in the training. Following is the training code to do this:

Listing 4.9 Training Neural Network with Dropout

```
def train(X, y, learning_rate, num_iterations, hidden_layer_
sizes, lambda_, dropout_rate):

    X = X.T
    Y = y.reshape(1, -1)
    nx, m = X.shape

    layer_sizes = [nx] + hidden_layer_sizes + [1]
    parameters = initialize_parameters(layer_sizes)

    costs = []

    for i_iter in range(num_iterations):
        A3, cache = forward_propagation(X, parameters, dropout_rate)

        cost = compute_cost(A3, Y, parameters, lambda_)

        grads = backward_propagation(A3, Y, lambda_, cache)

        parameters = update_parameters(parameters, grads,
learning_rate)

        costs.append(np.squeeze(cost))
        if i_iter % 400 == 0:
            print(f"Iteration: {i_iter}, Cost: {np.squeeze(cost)}")

    return parameters, costs
```

This code defines the training function for a neural network, incorporating dropout to prevent overfitting. During forward propagation, the `dropout_rate` parameter is used to randomly deactivate a fraction of neurons, ensuring the network does not rely too heavily on any single neuron. This helps the model generalize better to unseen data. The function also includes steps for initializing parameters, computing cost, performing backward propagation, and updating parameters iteratively.

Next, we will proceed with the training phase for a few iterations using a three-layer neural network, each layer consisting of 10 hidden nodes. We will not be applying regularization during the training phase. Let us first train without applying dropout:

```
num_iterations = 2000
learning_rate = 0.01
hidden_layer_sizes = [10, 10]
lambda_ = 0.0
dropout_rate = 0.2
parameters, costs = train(X_train, y_train, learning_rate,
num_iterations, hidden_layer_sizes, lambda_, dropout_rate)
```

This prints:

```
Iteration: 0, Cost: 0.7067701190821684
Iteration: 400, Cost: 0.47494388932731785
Iteration: 800, Cost: 0.3172323032374174
Iteration: 1200, Cost: 0.24658578444152157
Iteration: 1600, Cost: 0.2020990083416662
```

We will then train with dropout:

```
num_iterations = 2000
learning_rate = 0.01
hidden_layer_sizes = [10, 10]
lambda_ = 0.0
dropout_rate = 0.2
parameters, costs = train(X_train, y_train, learning_rate,
num_iterations, hidden_layer_sizes, lambda_, dropout_rate)
```

This prints:

```
Iteration: 0, Cost: 0.723071683947637
Iteration: 400, Cost: 0.5801189744478055
Iteration: 800, Cost: 0.4441529054704038
Iteration: 1200, Cost: 0.3535715655165295
Iteration: 1600, Cost: 0.2946437186465264
```

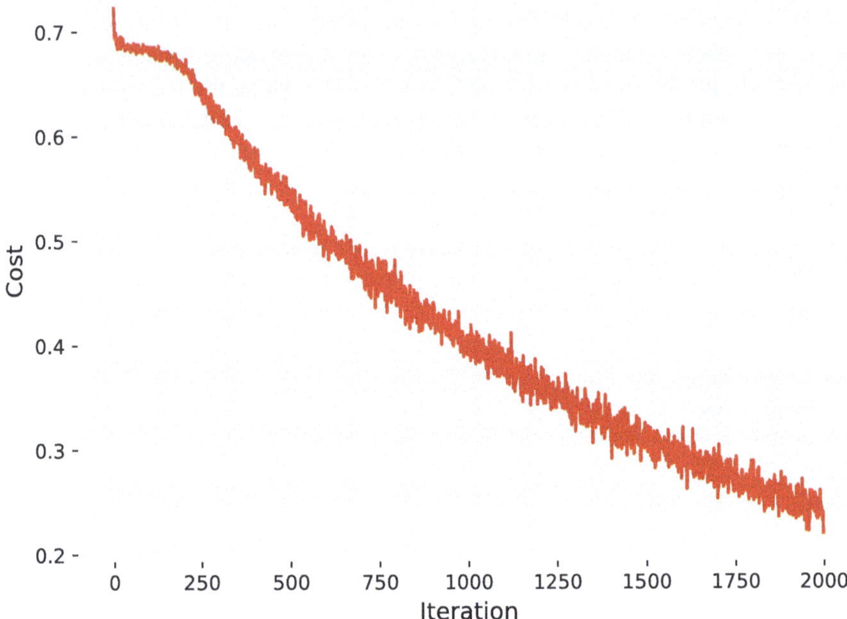

Fig. 4.4 The training cost curve using dropout. The training cost is expected to decrease over time

There is one interesting observation. For the same number of iterations, a model trained without dropout has a lower training loss compared to a model trained with dropout. This is because without dropout, the model can fully utilize all its neurons during training. This often results in lower training loss because the model can fit the training data more closely. When dropout is applied, some neurons are randomly dropped out during each training iteration. This introduces noise and can lead to higher training loss initially because the model has to learn to be more robust and generalize better.

We can also plot training cost curve as:

```
plot_costs(costs)
```

This would generate the training cost curve as shown in Fig. 4.4. We can see that the cost decreases over the iterations, indicating that the model has been trained correctly.

4.3.4 Prediction and Validation

As mentioned earlier, we do not apply dropout during the prediction phase. That is why we set the `dropout_rate` to zero during prediction. Here is the code for prediction:

Listing 4.10 Neural Network Prediction with Dropout

```
def predict(X, parameters, dropout_rate=0):

    X = X.T
    y_hat, _ = forward_propagation(X, parameters, dropout_rate)
    y_hat = np.squeeze(y_hat)

    return y_hat
```

Now, let us proceed to generate predictions using our test (validation) set and subsequently calculate the relevant classification metrics to evaluate the performance of our model:

```
y_hat = predict(X_test, parameters)
y_pred = (y_hat>0.5).astype(int)
precision, recall, f1 = classification_metrics(y_test, y_pred)
```

```
F1 Score: 0.8866995073891626
Precision: 0.9090909090909091
Recall: 0.8653846153846154
```

Without much tuning, we have attained an F1-score of 0.88 on the test (validation) set. Additionally, both our precision and recall metrics (Saito and Rehmsmeier 2015) exceeded 0.8. This signifies the successful training of our image classifier, which notably includes a dropout layer.

4.4 Early Stopping

The final technique we will explore to prevent overfitting is known as “early stopping” (Prechelt 2002). As we train a neural network for a number of iterations, we may wonder, when to stop our training? How long we should train for? Typically, we utilize a validation dataset and monitor the cost on this set as we train (Caruana et al. 2000). With each training iteration or epoch, we evaluate the model’s performance on this validation data. Figure 4.5 illustrates the concept. We stop training when loss stops improving on the validation set. This technique is known as early stopping.

This approach allows us to keep the model’s ability to generalize to new data, rather than fitting the model too closely to the training data. It is a kind of trade-off between bias and variance, aiming to find the sweet spot so that the model performs well on both the training data and unseen data.

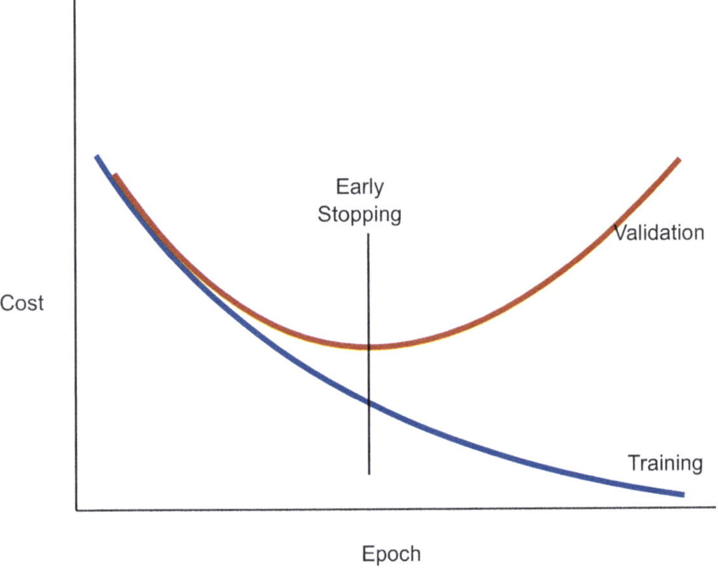

Fig. 4.5 Early stopping concept to prevent overfitting. Early stopping involves monitoring the model's performance on a validation set and halting the training process when the performance starts to degrade

4.5 Summary

- Bias and variance are crucial concepts in model performance, where bias indicates underfitting and variance indicates overfitting; balancing them is essential for optimal results.
- Regularization, such as L2 regularization, helps reduce overfitting by adding a penalty term to the cost function, controlling the complexity of the model.
- Dropout is a regularization technique that prevents overfitting by randomly deactivating a portion of neurons during training, promoting robust feature learning.
- Implementing regularization involves modifying the cost function and adjusting backward propagation to account for the regularization term, reducing the influence of large weights.
- Dropout implementation requires modifying forward propagation to apply dropout masks and adjusting backward propagation to ensure gradients are calculated correctly despite the dropped neurons.
- Practical applications, such as the KDD network intrusion detection and horses or humans image classification tasks, demonstrate the effectiveness of regularization and dropout in reducing model complexities.

References

Baldi, P., & Sadowski, P. J. (2013). Understanding dropout. *Advances in Neural Information Processing Systems, 26.*

Caruana, R., Lawrence, S., & Giles, C. (2000). Overfitting in neural nets: Backpropagation, conjugate gradient, and early stopping. *Advances in Neural Information Processing Systems, 13.*

Domingos, P. (2012). A few useful things to know about machine learning. *Communications of the ACM, 55*(10), 78–87.

Geman, S., Bienenstock, E., & Doursat, R. (1992). Neural networks and the bias/variance dilemma. *Neural Computation, 4*(1), 1–58.

Golub, G. H., & Van Loan, C. F. (2013). Matrix computations, 4th. *Johns Hopkins.*

Hastie, T., Tibshirani, R., Friedman, J., & others. (2009). *The elements of statistical learning.* Citeseer.

Hawkins, D. M. (2004). The problem of overfitting. *Journal of Chemical Information and Computer Sciences, 44*(1), 1–12.

Hoerl, A. E., & Kennard, R. W. (1970). Ridge regression: Biased estimation for nonorthogonal problems. *Technometrics, 12*(1), 55–67.

Krogh, A., & Hertz, J. (1991). A simple weight decay can improve generalization. *Advances in Neural Information Processing Systems, 4.*

Ng, A. Y. (2004). Feature selection, L 1 vs. L 2 regularization, and rotational invariance. *Proceedings of the Twenty-First International Conference on Machine Learning*, 78.

Pedregosa, F., Varoquaux, G., Gramfort, A., Michel, V., Thirion, B., Grisel, O., Blondel, M., Prettenhofer, P., Weiss, R., Dubourg, V., & others. (2011). Scikit-learn: Machine learning in Python. *The Journal of Machine Learning Research, 12*, 2825–2830.

Prechelt, L. (2002). Early stopping-but when? In *Neural Networks: Tricks of the trade* (pp. 55–69). Springer.

Saito, T., & Rehmsmeier, M. (2015). The precision-recall plot is more informative than the ROC plot when evaluating binary classifiers on imbalanced datasets. *PLoS One, 10*(3), e0118432.

Srivastava, N., Hinton, G., Krizhevsky, A., Sutskever, I., & Salakhutdinov, R. (2014). Dropout: A simple way to prevent neural networks from overfitting. *The Journal of Machine Learning Research, 15*(1), 1929–1958.

Tavallaee, M., Bagheri, E., Lu, W., & Ghorbani, A. A. (2009). A detailed analysis of the KDD CUP 99 data set. *2009 IEEE Symposium on Computational Intelligence for Security and Defense Applications*, 1–6.

Van Rijsbergen, C. J. (1979). *Information Retrieval, 2nd edn. Newton, MA.* USA: Butterworth-Heinemann.

Chapter 5
Leveraging Advanced Optimization Techniques

This chapter covers:

- Understanding batch sizes, epochs, and iterations in gradient descent
- Exploring momentum-based gradient descent
- Applying learning rate decay techniques
- Demonstrating the application of advanced optimization techniques
- Introducing batch normalization for enhancing the stability of neural networks

In the previous chapters, we have laid the groundwork by discussing fundamental optimization techniques essential for training deep learning models. However, the basic optimization techniques may fall short in certain scenarios and might lead to slow convergence or suboptimal performance. Now, we dive into more sophisticated strategies to overcome these challenges. This chapter explores advanced optimization techniques, beginning with different variants of gradient descent methods tailored to various batch sizes. As datasets continue to grow, full batch gradient descent become impractical due to the computational cost associated with processing the entire dataset in each iteration. Therefore, understanding the nuances of mini-batch and stochastic gradient descent becomes crucial. From there, we transition to momentum-based methods designed to overcome the limitations of simple gradient descent, such as slow convergence and susceptibility to local minima. We will also cover advanced algorithms such as RMSprop and Adam, which integrate momentum and adaptive learning rates to further accelerate training and improve model accuracy. We will look at practical examples and code implementations. This chapter will equip us with the tools to leverage these advanced techniques, ensuring our deep learning models are both efficient and effective.

T. Islam, *Hands-on Deep Learning*, https://doi.org/10.1007/978-3-032-00488-8_5

5.1 Gradient Descent Methods and Batch Size

In our discussions thus far, we have been using all the examples in our training set in a single iteration. This means our batch size is equal to the total number of training examples. This approach is commonly known as full batch gradient descent. However, in modern machine learning, where datasets are often too large to fit into memory entirely, employing full batch gradient descent becomes impractical. Therefore, alternative variations of the gradient descent algorithm are used to accommodate different batch sizes (Bottou, 2010; Li et al., 2014). Let us explore these further.

There are three gradient descent methods related to batch size:

- *Full batch gradient descent*: The entire set of training examples is used in each step/iteration/pass for training. This is also known as batch gradient descent. The batch size is equal to m. If training set is small, batch gradient descent is recommended.
- *Mini batch gradient descent*: A subset of training examples is used in each step/iteration/pass. Generally, batch sizes of 32, 64, 128, 256, 512 are used. This means, with a batch size of 64, 64 examples are used in each step/iteration/pass. If training set is medium to large, mini batch gradient descent is recommended. Most deep learning applications fit in this category.
- *Stochastic gradient descent*: In this method, a single training example is used in each step/iteration/pass. This means the batch size is 1. This is recommended for online learning.

There is another terminology we need to remember related to batch size: **epoch**. A single epoch means that we have completed one pass of all training examples (Fig. 5.1). This means, in mini batch gradient descent, 1 epoch consists of several mini batches or steps/iterations. The number of passes over the training set in an epoch depends on the batch size and the total number of training examples.

Let us look at an example as we are training a model in TensorFlow (Fig. 5.2). Consider we have 3060 samples in our training set. Note that we are using a batch

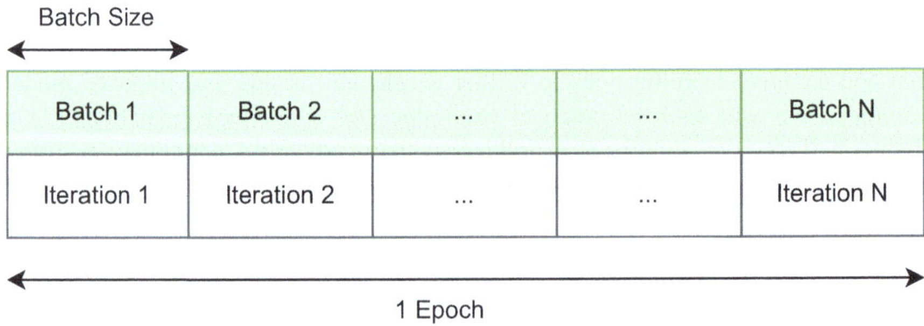

Fig. 5.1 The relationship between batch size, iteration, and epoch. A batch size is a subset of data used in one training step, an iteration is one pass through a batch, and an epoch is one complete pass through the entire dataset

```
Epoch 2/5
96/96 [==============================] - 2s 20ms/step - loss: 4.3675 - accuracy: 0.0529 - val_loss: 4.2558 - val_accuracy: 0.0653
Epoch 3/5
96/96 [==============================] - 2s 19ms/step - loss: 3.8272 - accuracy: 0.1474 - val_loss: 3.3359 - val_accuracy: 0.3070
Epoch 4/5
96/96 [==============================] - 2s 19ms/step - loss: 3.3319 - accuracy: 0.2399 - val_loss: 2.9436 - val_accuracy: 0.3466
Epoch 5/5
96/96 [==============================] - 2s 19ms/step - loss: 2.9639 - accuracy: 0.3095 - val_loss: 3.0435 - val_accuracy: 0.3529
```

Fig. 5.2 An example illustrating the concept of epoch. An epoch represents one complete pass through the entire dataset

size of 32 in this example. We can see that each epoch consists 96 total number of batches. This implies that there will be 96 iterations in each epoch. In this case, "3060 samples/32 samples per batch = 95.625 batches." Since we cannot have a fraction of a batch, TensorFlow processes 95 full batches of 32 samples and 1 additional batch with the remaining 20 samples, for a total of 96 batches.

Below is the code that implements the creation of mini batches, given the entire training set and batch size. Here, we are shuffling our datasets first. This removes the ordering in the data and ensures the model is exposed to diverse types of data in an iteration. This leads to a better generalization. Then, we are appending each batch with 64 (for `batch_size` = 64) examples to the `mini_batches` list. This `mini_batches` can then be used to train a neural network model in an iterative manner.

Listing 5.1 Function to Create Mini Batches

```
import numpy as np

def create_mini_batches(X, y, batch_size, seed):

    np.random.seed(seed)

    indices = np.arange(X.shape[0])
    np.random.shuffle(indices)

    mini_batches = []
    for start_idx in range(0, X.shape[0] - batch_size +
1, batch_size):
        excerpt = indices[start_idx:start_idx + batch_size]
        mini_batches.append((X[excerpt], y[excerpt]))
    return mini_batches
```

We can generate list of batches by calling the above function. For instance, we can create mini batches with a batch size of 64 like this:

```
batch_size = 64
mini_batches = create_mini_batches(X_train, y_train, batch_size)
```

5.2 Momentum-Based Gradient Descent

In the second chapter, we looked into the workings of gradient descent algorithm. This algorithm, while crucial for grasping optimization in deep learning, does come with its own set of limitations. Specifically, (1) it can become trapped in local minima or saddle points (Dauphin et al., 2014) on the error surface, which are not the optimal solutions, and (2) its convergence speed can be significantly influenced by the choice of learning rate, with an inappropriate rate leading to slower convergence or even divergence. This is where the momentum-based gradient descent (Polyak, 1964) comes in. It addresses the limitations of the basic gradient descent algorithm by avoiding local minima and improving convergence speed.

Momentum-based gradient descent is a method that helps accelerate gradients vectors in the right directions, thus leading to faster convergence (Polyak, 1964; Qian, 1999; Sutskever et al., 2013). It is among the most commonly used optimization algorithms and is extensively employed in the training of deep learning models.

Let us look at an example objective function that has multiple local minimas:

$$x^2 + \sin(2\pi x)$$

The following code implements both the simple gradient descent and the momentum-based gradient descent algorithms for the above objective function. In this code, we have defined the objective function first. The derivative of the objective function is also defined in `df(x)` function for gradient calculation. In momentum-based gradient descent, a velocity term is introduced to accelerate the gradient descent in the relevant direction and to dampen the oscillations. This velocity term is a moving average over the past gradients. It essentially adds a proportion of the previous update to the current one, thereby gaining momentum and speeding up the convergence toward the minimum:

Listing 5.2 Functions for Simple and Momentum-Based Gradient Descents

```
def f(x):
    return x**2 + np.sin(2 * np.pi * x)

def df(x):
    return 2 * x + 2 * np.pi * np.cos(2 * np.pi * x)

def simple_gradient_descent(start_at, learning_rate,
num_iterations):
    positions = np.zeros(num_iterations + 1)
    positions[0] = start_at
    for i in range(num_iterations):
```

(continued)

Listing 5.2 (continued)

```
        gradient = df(positions[i])
        positions[i + 1] = positions[i] - learning_rate * gradient
    return positions

def momentum_based_gradient_descent(start_at, learning_rate,
momentum, num_iterations):
    positions = np.zeros(num_iterations + 1)
    velocity = 0
    positions[0] = start_at
    for i in range(num_iterations):
        gradient = df(positions[i])
        velocity = momentum * velocity + gradient
        positions[i + 1] = positions[i] - learning_rate * velocity

    return positions
```

Let us now apply the two optimization methods for few iterations and plot them:

```
start_at = 2.0
learning_rate = 0.01
momentum = 0.9
num_iterations = 100

positions_momentum = gradient_descent_with_momentum(start_at,
learning_rate, momentum, num_iterations)

positions_simple = simple_gradient_descent(start_at, learning_rate,
num_iterations)

plt.figure()
plt.subplot(2, 1, 1)
x = np.linspace(-2.0, 2.0, 400)
y = f(x)
plt.plot(x, y, alpha=0.6, label='Objective function')
plt.scatter(positions_simple, f(positions_simple), color='blue',
label='Optimizer position')
plt.title('Simple gradient descent')
plt.xlabel('x')
plt.ylabel('f(x)')
plt.legend()
plt.tight_layout()
```

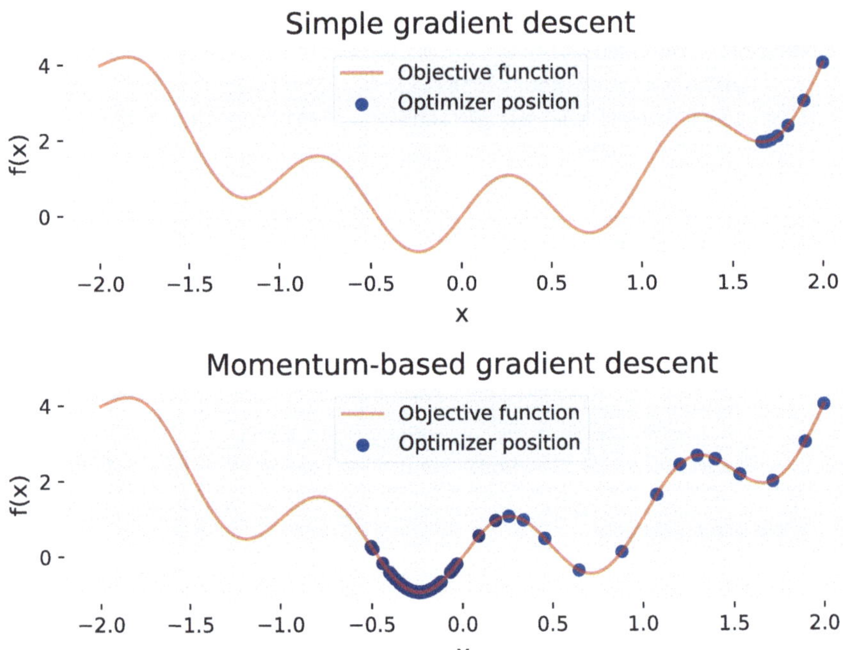

Fig. 5.3 A comparative illustration demonstrating how momentum-based gradient descent aids in escaping local minima

```
plt.subplot(2, 1, 2)
x = np.linspace(-2.0, 2.0, 400)
y = f(x)
plt.plot(x, y, alpha=0.6, label='Objective function')
plt.scatter(positions_momentum, f(positions_momentum),
color='blue', label='Optimizer position')
plt.title('Momentum-based gradient descent')
plt.xlabel('x')
plt.ylabel('f(x)')
plt.legend()
plt.tight_layout()
```

The execution of the above code will result in the comparison plot shown in Fig. 5.3. The objective function has multiple minima. We can see that the simple gradient descent algorithm gets stuck in a local minimum. However, the momentum-based gradient descent algorithm has been able to escape the local minima and find the global minimum.

There are few popular momentum-based optimization algorithms to update the weights in neural networks. We will discuss them next.

5.2.1 *Gradient Descent with Momentum*

We have just covered the rationale of using momentum-based gradient descent method. This technique can be particularly effective in navigating the complex landscape of the dataset. By incorporating momentum, the algorithm can build up velocity and push through regions of low gradient, potentially avoiding getting trapped in suboptimal solutions. This means that instead of using the regular gradient descent algorithm, we can achieve better results by employing gradient descent with momentum.

Let us now discuss how the parameters are updated in this algorithm. The formulas for gradient descent with momentum will be:

1. Compute the gradient of the loss function with respect to parameters:

$$\nabla_{W,b} J(W, b)$$

2. Update the velocity v, which is the moving average of the negative gradient:

$$v = \beta \cdot v + (1 - \beta) \cdot \nabla_{W,b} J(W, b)$$

3. Update the parameters with the velocity:

$$W = W - \alpha \cdot v$$

$$b = b - \alpha \cdot v$$

where:

- W and b represent the weights and biases of the neural network.
- $J(W, b)$ is the cost function.
- $\nabla_{W,\, b} J(W, b)$ is the gradient of the cost function.
- α is the learning rate.
- β is the momentum term (usually set to 0.9).
- v is the velocity (initialized at 0).

This way, the gradient descent with momentum is applied to update weights and biases of the neural network. The velocity is fundamentally a weighted average of the gradients, introducing momentum into the update process, hence the method's name. Momentum considers previous gradients to smooth out the update. We calculate an exponentially weighted average of the gradients and use this averaged gradient for updating the weights. This can help to accelerate learning and overcome local minima. The parameter updates then take a step in the direction of the velocity. The learning rate α controls the size of this step.

To be consistent with notations, let us emphasize the parameters update for the respective layers l. We know from the previous chapter that, in regular gradient descent, the weights and bias parameters are updated as follows:

$$dW^{[l]} = dZ^{[l]}A^{[l-1]}$$

$$db^{[l]} = dZ^{[l]}$$

$$W^{[l]} = W^{[l]} - \alpha dW^{[l]}$$

$$b^{[l]} = b^{[l]} - \alpha db^{[l]}$$

In the context of gradient descent with momentum, the weights and bias parameters are updated as follows:

Weights update:

$$v_{dW^{[l]}} = \beta v_{dW^{[l]}} + (1 - \beta)dW^{[l]}$$

$$W^{[l]} = W^{[l]} - \alpha v_{dW^{[l]}}$$

Bias update:

$$v_{db^{[l]}} = \beta v_{db^{[l]}} + (1 - \beta)db^{[l]}$$

$$b^{[l]} = b^{[l]} - \alpha v_{db^{[l]}}$$

A Python implementation of gradient descent with momentum will look like this:

```
import numpy as np

W = np.random.randn(4, 3)
dW = np.random.randn(4, 3)

num_iterations = 10
v_dW = 0
v_dB = 0
beta = 0.9
alpha = 0.01
for t in range(num_iterations):
    v_dW = (beta * v_dW) + (1 - beta) * dW
    W = W - alpha * v_dW
```

5.2.2 *Gradient Descent with RMSprop*

The RMSprop is another variant of momentum-based gradient descent algorithm that will speed up the learning process (Tieleman & Hinton, 2012; Xu et al., 2021). The mathematical formulas to update the $W^{[l]}$ and $b^{[l]}$ parameters are as follows:

Weights update:

$$s_{dW^{[l]}} = \beta s_{dW^{[l]}} + (1 - \beta)\left(dW^{[l]}\right)^2$$

$$W^{[l]} = W^{[l]} - \alpha \frac{dW^{[l]}}{\sqrt{s_{dW^{[l]}}} + \epsilon}$$

Bias update:

$$s_{db^{[l]}} = \beta s_{db^{[l]}} + (1 - \beta)\left(db^{[l]}\right)^2$$

$$b^{[l]} = b^{[l]} - \alpha \frac{db^{[l]}}{\sqrt{s_{db^{[l]}}} + \epsilon}$$

where:

- β is the decay rate. This is a new hyperparameter. A common choice of $\beta = 0.9$.
- α is the learning rate.
- ϵ is a very small number to prevent any division by zero in the update. A common choice of $\epsilon = 10^{-8}$.

5.2.3 *Gradient Descent with Adam*

Adam or adaptive momentum is another optimization algorithm that combines the concepts of gradient descent with momentum and RMSprop (Kingma, 2014). It has become a go-to choice for deep learning training. The mathematical formulas to update the $W^{[l]}$ and $b^{[l]}$ parameters are as follows, where t represents the current iteration number used for bias correction:

$$v_{W^{[l]}} = \beta_1 v_{W^{[l]}} + (1 - \beta_1)\frac{\partial J}{\partial W^{[l]}}$$

$$v_{W^{[l]}}^{corrected} = \frac{v_{W^{[l]}}}{1 - (\beta_1)^t}$$

$$s_{W^{[l]}} - \beta_2 s_{W^{[l]}} + (1 - \beta_2)\left(\frac{\partial J}{\partial W^{[l]}}\right)^2$$

$$s_{W^{[l]}}^{corrected} = \frac{s_{W^{[l]}}}{1 - (\beta_2)^t}$$

$$W^{[l]} = W^{[l]} - \alpha \frac{v_{W^{[l]}}^{corrected}}{\sqrt{s_{W^{[l]}}^{corrected}} + \varepsilon}$$

The hyperparameters for Adam optimization and their suggested values are provided in the table as follows:

Adam hyperparameters	Suggested value
alpha	Tune
beta1	0.9
beta2	0.999
epsilon	10^{-8}

The code snippet provided below is the implementation of the Adam optimization algorithm. This algorithm begins by initializing the first-order moment (denoted as v) and the second-order moment s. Following this initialization, the algorithm proceeds to update the weights and biases:

Listing 5.3 Adam Optimization Algorithm

```
def initialize_moments(parameters):
    moments = {}
    for key in parameters.keys():
        moments["v" + key] = np.zeros_like(parameters[key])
        moments["s" + key] = np.zeros_like(parameters[key])
    return moments

def update_parameters_adam(
    parameters, grads, moments, t, learning_rate=0.01, beta1=0.9,
beta2=0.999, epsilon=1e-8
):
    for key in parameters.keys():
        moments["v" + key] = beta1 * moments["v" + key] + (1 - beta1) *
grads["d" + key]
        moments["s" + key] = beta2 * moments["s" + key] + (1 - beta2) *
np.square(
            grads["d" + key]
        )

        v_corrected = moments["v" + key] / (1 - beta1**t)
        s_corrected = moments["s" + key] / (1 - beta2**t)

        parameters[key] -= (
            learning_rate * v_corrected / (np.sqrt(s_corrected) +
epsilon)
        )
    return parameters, moments
```

In order to execute the above code, we will first need to initialize our parameters with He initialization. The following snippet shows how we can do this. We are assuming a three-layer neural network:

```
hidden_layer_sizes = [5, 3]
nx = 10
layer_sizes = [nx] + hidden_layer_sizes + [1]
parameters = initialize_parameters(layer_sizes)
print (parameters.keys())
```

This will print the following keys representing the parameters for the three layers:

```
dict_keys(['W1', 'b1', 'W2', 'b2', 'W3', 'b3'])
```

Now, we can initialize moments:

```
moments = initialize_moments(parameters)
print (moments.keys())
```

This gives us the moments for the corresponding weights and biases:

```
dict_keys(['vW1', 'sW1', 'vb1', 'sb1', 'vW2', 'sW2', 'vb2', 'sb2',
'vW3', 'sW3', 'vb3', 'sb3'])
```

5.3 Learning Rate Decay

In the context of gradient descent (with or without moments) optimization, a fixed learning rate can often lead to a solution that is somewhat distant from the optimal convergence point. This is where the concept of learning rate decay comes into play.

Learning rate decay is a strategy that involves gradually reducing the learning rate as the training process progresses (You et al., 2019; Zeiler, 2012). This approach offers two key advantages:

- *Acceleration of initial training*: In the early stages of training, a higher learning rate is used, which allows the model to take larger steps in the parameter space and thus speeds up the initial learning process.
- *Precision in convergence*: As the training process approaches convergence, the learning rate is reduced. This results in smaller steps in the parameter space, allowing the model to fine-tune its parameters and converge more precisely to the optimal solution.

In essence, learning rate decay provides a balance between speed and precision in the training process, enabling the model to converge more accurately without significantly compromising on the speed of training.

There are various forms of learning rate decay used by deep learning practitioners (Bengio, 2012). For example,

- Inverse time decay
- Exponential decay

5.3.1 Inverse Time Decay

Here is the formula for inverse time decay in a more mathematical representation:

$$lr_{epoch} = \frac{1}{1 + decay \times epoch} \times lr_{initial}$$

A Python implementation for inverse time decay will look like this:

```
initial_learning_rate = 0.2
decay_rate = 1
num_epochs = 10
for epoch in range(num_epochs):
    learning_rate = (1/(1+ (decay_rate * epoch))) *
initial_learning_rate
    print(f"epoch: {epoch}, learning_rate: {learning_rate}")
```

Executing above code will result in following learning rates:

```
epoch: 0, learning_rate: 0.01
epoch: 1, learning_rate: 0.005
epoch: 2, learning_rate: 0.0033
epoch: 3, learning_rate: 0.0025
epoch: 4, learning_rate: 0.002
epoch: 5, learning_rate: 0.0017
epoch: 6, learning_rate: 0.0014
epoch: 7, learning_rate: 0.0012
epoch: 8, learning_rate: 0.0011
epoch: 9, learning_rate: 0.001
```

5.3.2 Exponential Decay

Here is the formula for inverse time decay in a more mathematical representation:

$$lr_{epoch} = \frac{1}{1 + decay \times epoch} \times lr_{initial}$$

A Python implementation for inverse time decay will look like this:

```
import numpy as np
initial_learning_rate = 0.1
decay_rate = 0.5
num_epochs = 10
for epoch in range(num_epochs):
    learning_rate = initial_learning_rate * np.exp(-decay_rate * epoch)
    print(f"epoch: {epoch}, learning_rate: {np.round(learning_rate,
4)}")
```

Executing above code will result in following learning rates:

```
epoch: 0, learning_rate: 0.1
epoch: 1, learning_rate: 0.0607
epoch: 2, learning_rate: 0.0368
epoch: 3, learning_rate: 0.0223
epoch: 4, learning_rate: 0.0135
epoch: 5, learning_rate: 0.0082
epoch: 6, learning_rate: 0.005
epoch: 7, learning_rate: 0.003
epoch: 8, learning_rate: 0.0018
epoch: 9, learning_rate: 0.0011
```

5.4 Training with Advanced Optimization Techniques

5.4.1 Diamonds Price Prediction Problem

In this chapter, we will work with diamonds price dataset to develop intuitive understanding on advanced optimization techniques by applying the techniques discussed earlier. This dataset contains physical attributes and prices of diamonds. The attributes include characteristics of diamonds such as weight, cut, color, clarity, dimensions, and so on. Given these attributes, the task is to predict price of the diamond.

First, we will load the data from TensorFlow datasets. Here is the loading function:

Listing 5.4 Loading Diamonds Data from TensorFlow Datasets

```
import numpy as np
import tensorflow_datasets as tfds

def load_diamonds_data():

    dataset, metadata = tfds.load('diamonds', with_info=True,
split='train')

    X = []
    y = []

    for example in dataset:
        y.append(example['price'])
        X.append(list(example['features'].values()))

    X = np.array(X)
    y = np.array(y)

    return X, y
```

We can then run:

```
X, y = load_diamonds_data()
```

Next, we will divide the dataset into training and validation subsets. Additionally, we will normalize the input features:

```
X_train, X_test, y_train, y_test = split_dataset(X, y)
X_train, X_test = normalize_features(X_train, X_test)
```

5.4.2 Training

It is time to train a neural network leveraging advanced optimization techniques. We are predicting the price of diamonds. This means we are dealing with a regression problem here. It is important to note that the distribution of diamond prices is skewed, as illustrated in the histogram plot in Fig. 5.4. In regression problems, it is often beneficial to transform a skewed target variable to approximate a Gaussian

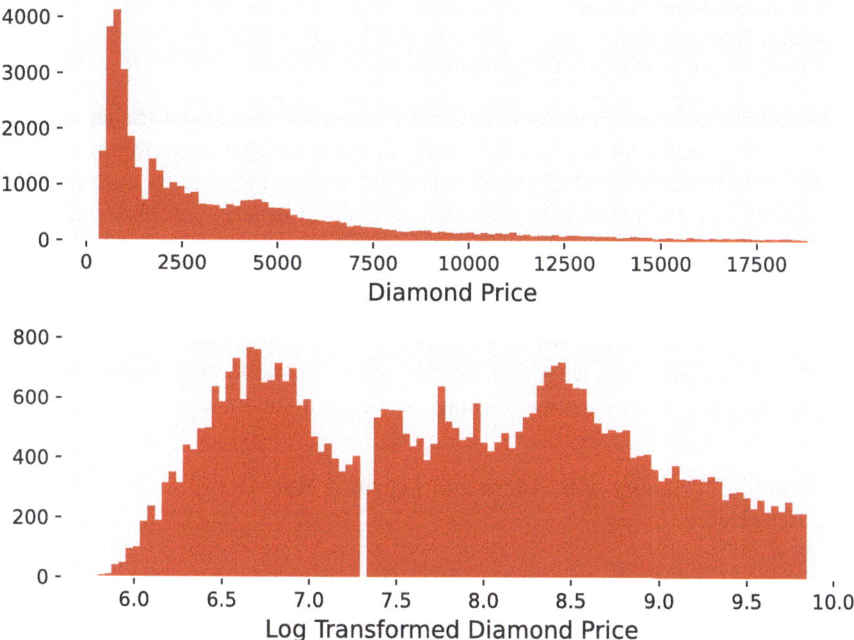

Fig. 5.4 Illustration of the distribution of diamond prices before and after applying a log transformation, highlighting the effect of the transformation on the data distribution

distribution (Osborne, 2010). This transformation can lead to improved model performance for several reasons:

- *Stabilized training process*: Neural networks can be sensitive to the scale and distribution of the target variable. Transformation can stabilize the training process, making it more robust and less prone to overfitting.
- *Outlier handling*: The transformation can also mitigate the influence of outliers, leading to more robust and generalizable models.

A common choice for such a transformation is the logarithmic transformation. For instance, the target variable `y_train` can be transformed as:

```
y_train = np.log1p(y_train)
```

Next, we define the cost function for the regression problem:

$$J = \frac{1}{2m}\sum_{i=1}^{m}(\hat{y_i} - y_i)^2$$

where:

- $\hat{y}_i$ is the predicted output
- y_i is the actual output
- m is the number of examples in the mini batch

This above equation represents the mean squared error (MSE) (Seber & Lee, 2003) cost function. Our goal is to minimize this cost function J leveraging optimization algorithm such as Adam. The factor of 1/2 is often included for convenience, as it cancels out when taking the derivative in gradient descent optimization.

Next, we implement the training function to train the neural network. The training function incorporates optimization with regular gradient descent and Adam for comparison. Besides, we have incorporated inverse time decay in the training process. Using inverse time decay with Adam is not very common because Adam already has an adaptive learning rate. However, the following implementation demonstrates how to incorporate learning rate decay in the training process as well:

Listing 5.5 Training with Adam and Learning Rate Decay

```
def train(
    X,
    y,
    initial_learning_rate,
    decay_rate,
    num_epochs,
    hidden_layer_sizes,
    batch_size,
    optimizer,
):

    nx = X.shape[1]
    layer_sizes = [nx] + hidden_layer_sizes + [1]
    parameters = initialize_parameters(layer_sizes)

    moments = initialize_moments(parameters)

    costs = []
    t = 1

    for i_epoch in range(num_epochs):
        mini_batches = create_mini_batches(X, y, batch_size,
seed=i_epoch)

        costs_i_epoch = []
        for mini_batch in mini_batches:
            X_minibatch, y_minibatch = mini_batch
```

(continued)

Listing 5.5 (continued)

```
            X_minibatch = X_minibatch.T
            Y_minibatch = y_minibatch.reshape(1, -1)

            A3, cache = forward_propagation(X_minibatch,
parameters)

            cost = compute_cost(A3, Y_minibatch)

            grads = backward_propagation(A3, Y_minibatch, cache)

            learning_rate = (1 / (1 + (decay_rate * i_epoch))) *
initial_learning_rate

            if optimizer == "gradient_descent":
                parameters = update_parameters_gd(parameters,
grads, learning_rate)
            elif optimizer == "adam":
                parameters, moments = update_parameters_adam(
                    parameters,
                    grads,
                    moments,
                    t,
                    learning_rate,
                    beta1=0.9,
                    beta2=0.999,
                    epsilon=1e-8,
                )
                t = t + 1

            costs_i_epoch.append(np.squeeze(cost))

        costs.append(np.mean(costs_i_epoch))
        if i_epoch % 20 == 0:
            print(f"Epoch: {i_epoch}, Cost: {np.squeeze(np.mean
(costs_i_epoch))}")

    return parameters, costs
```

Now we have the training function. Let us train a three-layer neural network for the diamonds price prediction using the two optimization techniques, the regular gradient descent, and the Adam optimization:

```
num_epochs = 100
initial_learning_rate = 0.01
decay_rate = 0.01
hidden_layer_sizes = [5, 3]
batch_size = 64

optimizer = "gradient_descent"
parameters, costs_gd = train(X_train, y_train,
initial_learning_rate, decay_rate, num_epochs, hidden_layer_sizes,
batch_size, optimizer)

optimizer = "adam"
parameters, costs_adam = train(X_train, y_train,
initial_learning_rate, decay_rate, num_epochs, hidden_layer_sizes,
batch_size, optimizer)
```

We can now compare the training cost for the two optimization algorithms with the following code snippet:

```
plt.figure()
plt.plot(costs_gd)
plt.plot(costs_adam)
plt.xlabel("Epoch")
plt.ylabel("Training Cost")
plt.yscale('log')
plt.legend(['Simple GD', 'Adam'])
plt.xlim([0, 100])
```

The code provided above generates the training cost plot depicted in Fig. 5.5. It becomes evident that the Adam optimization algorithm achieves convergence more rapidly than the standard gradient descent algorithm when the number of epochs is held constant. This demonstrates the efficiency of Adam optimization for our use case.

5.4.3 *Prediction and Validation*

Advanced optimization techniques, such as Adam, are primarily applied during the training phase of a deep learning model. Once the model is trained, the prediction process remains unchanged, relying on the learned parameters to make predictions on new data.

Next, we will implement the predict function with the trained model. The following code implements the predict function, which essentially performs forward

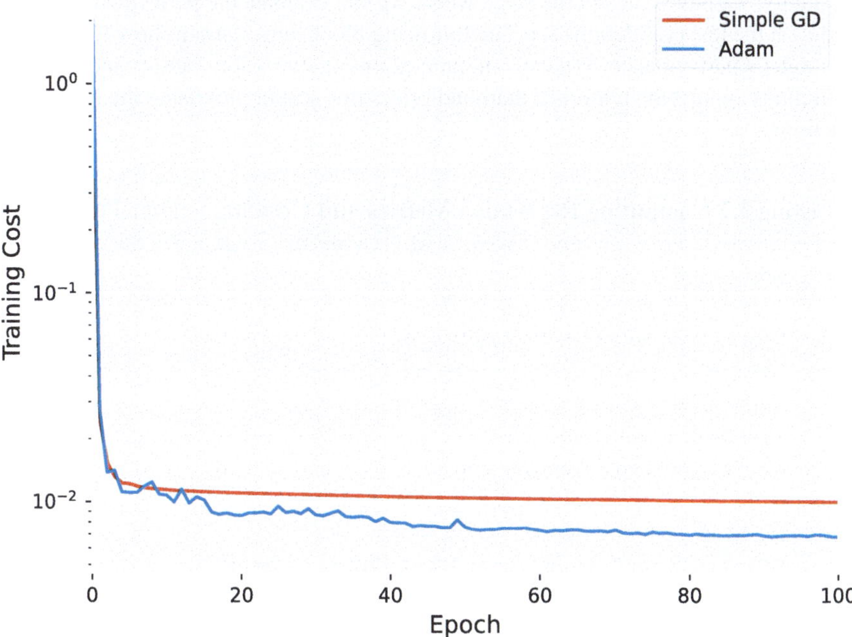

Fig. 5.5 Illustrating the convergence speed of cost minimization using the Adam optimization algorithm over traditional gradient descent

propagation using the trained parameters. It is noteworthy that to revert to the original diamond price from the transformed predicted price, we must reverse the log transformation. For this purpose, we utilize the `expm1` function:

Listing 5.6 Neural Network Prediction with Reverse Log Transformation

```
def predict(X, parameters):

    X = X.T
    y_hat, _ = forward_propagation(X, parameters)
    y_hat = np.squeeze(y_hat)

    y_hat = np.expm1(y_hat)

    return y_hat
```

```
y_hat = predict(X_test, parameters)
```

Finally, we have reached the stage where we can evaluate the performance of our model on the test (validation) set. The following block demonstrates how to validate the model's performance. We can compute essential regression metrics and plot the predictions against the reference diamond prices in a scatter plot using the following functions:

Listing 5.7 Computing Regression Metrics and Creating Scatter Plots

```
from sklearn.metrics import mean_absolute_error, mean_squared_error,
r2_score

def regression_metrics(y_test, y_hat):
    mae = mean_absolute_error(y_test, y_hat)
    mse = mean_squared_error(y_test, y_hat)
    rmse = np.sqrt(mse)
    r2 = r2_score(y_test, y_hat)

    print(f"Mean Absolute Error (MAE): {mae}")
    print(f"Root Mean Squared Error (RMSE): {rmse}")
    print(f"R-Squared: {r2}")

    return mae, rmse, r2

def plot_predictions(y_test, y_hat):

    plt.scatter(y_test, y_hat)
    plt.xlim([0, 20000])
    plt.ylim([0, 20000])
    plt.xlabel("Reference Diamond Price")
    plt.ylabel("Predicted Diamond Price")

    return None
```

By executing the above functions with the following code, we generate the scatter plot depicted in Fig. 5.6, along with the output of the regression metrics:

```
plot_predictions(y_test, y_hat)
mae, rmse, r2 = regression_metrics(y_hat, y_test)

Mean Absolute Error (MAE): 332.57026946002145
Root Mean Squared Error (RMSE): 691.2069958836596
R-Squared: 0.9692741220784252
```

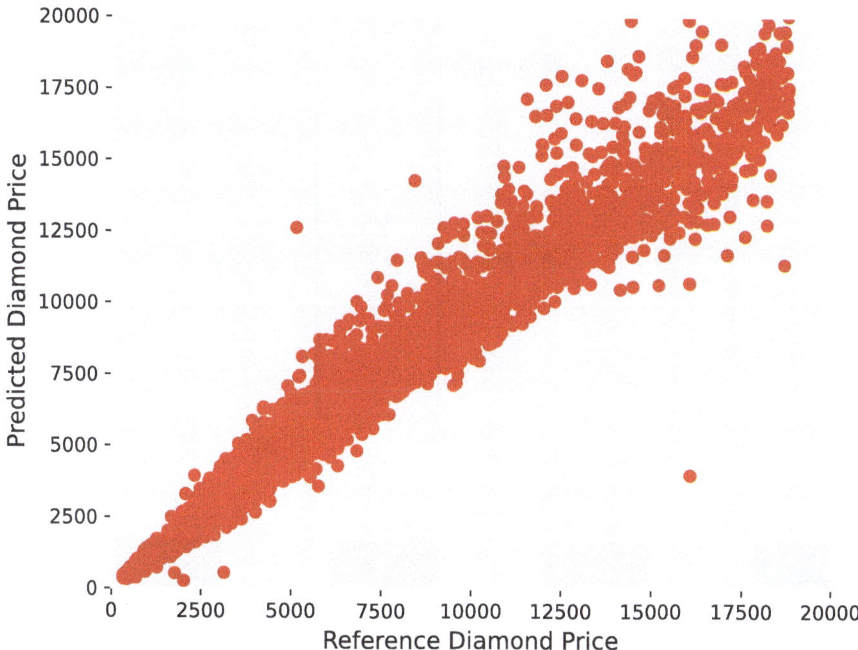

Fig. 5.6 The scatter plot visualizes the relationship between predicted diamond prices (generated by the deep learning model) and corresponding reference values (actual or known prices)

This is great! We have successfully implemented a comprehensive diamond price prediction model, utilizing advanced optimization techniques such as Adam optimization and learning rate decay. The root mean square error (RMSE) (Willmott & Matsuura, 2005) of our model is less than US$ 700.

5.5 Batch Normalization

There is one more important concept that we have not covered yet related to optimization—batch normalization (Ioffe & Szegedy, 2015). This technique is useful in enhancing the performance and stability of neural networks.

Previously, we discussed the normalization of input features. The main idea for input normalization is to adjust the inputs to a standard scale, making the mean of the input feature matrix 0 and the variance 1. This can help gradient descent to converge more quickly toward the minimum. It can also improve numerical stability.

Batch normalization extends the idea of normalization to the intermediate layers of the network, not just the input. More specifically, it normalizes the pre-activation weighted sum matrix Z, or the inputs to the activation function in each layer. In addition, batch normalization introduces two learnable parameters. These parameters, often denoted as gamma (scale) and beta (shift), are learned through the same

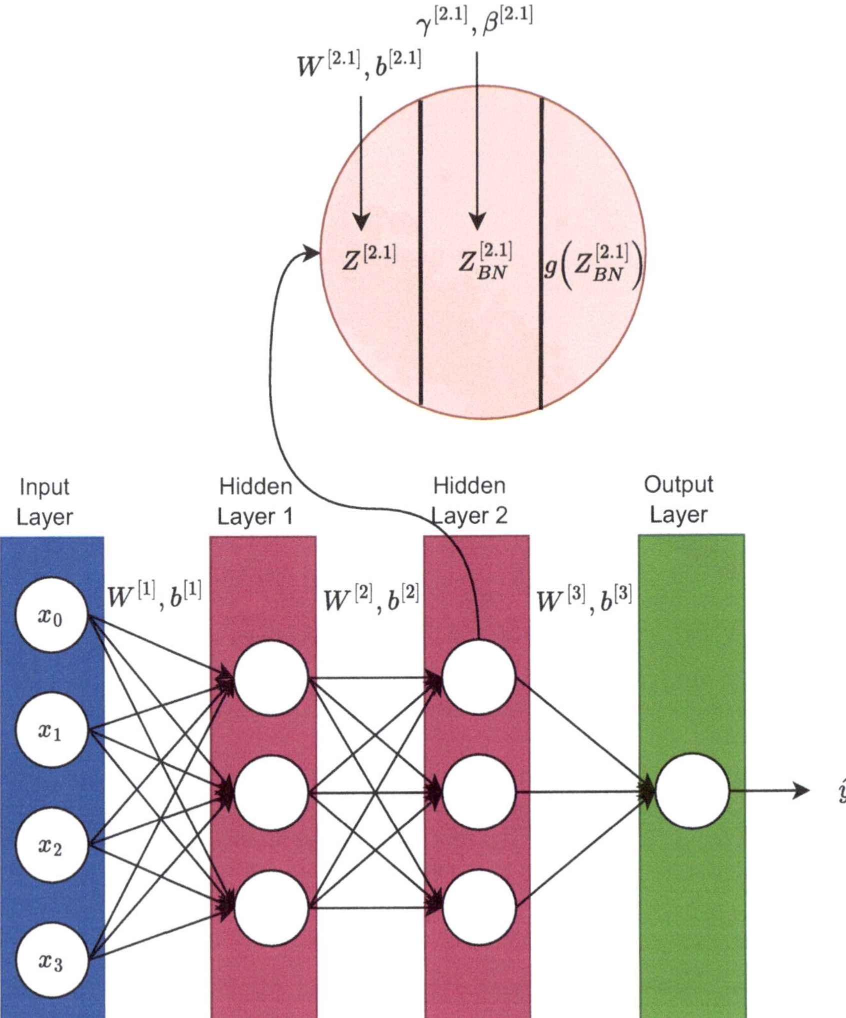

Fig. 5.7 Illustration of batch normalization. Batch normalization is a technique that standardizes the inputs to a layer

gradient descent process as the other parameters in the network. This is because instead of enforcing a strict zero mean and unit variance, the network has the flexibility to learn the optimal mean and variance for each layer's inputs. The primary purpose of using batch normalization is to speed up the learning and improve numerical stability (Santurkar et al., 2018).

The batch normalization equations can be conceptualized in Fig. 5.7 for a single activation node. In particular, batch normalization can be thought of an intermediate step between the computation of Z and A matrices.

In the standard forward propagation, we have:

$$Z^{[l]} = W^{[l]}A^{[l-1]} + b^{[l]}$$

$$A^{[l]} = g^{[l]}\left(Z^{[l]}\right)$$

where:

- $Z^{[l]}$ is the linear output of layer l
- $W^{[l]}$ and $b^{[l]}$ are the weight and bias of layer l
- $A^{[l-1]}$ is the activation output from the previous layer l
- $g^{[l]}$ is the activation function of layer l

In batch normalization, we normalize $Z^{[l]}$ before applying the activation function. The equations become:

$$Z^{[l]} = W^{[l]}A^{[l-1]} + b^{[l]}$$

$$\mu^{[l]} = \frac{1}{m}\sum\nolimits_{i=1}^{m} Z_i^{[l]}$$

$$\sigma^{[l]} = \sqrt{\frac{1}{m}\sum\nolimits_{i=1}^{m}\left(Z_i^{[l]} - \mu^{[l]}\right)^2}$$

$$Z_{norm}^{[l]} = \frac{Z^{[l]} - \mu^{[l]}}{\sigma^{[l]}}$$

$$Z_{BN}^{[l]} = \gamma^{[l]} Z_{norm}^{[l]} + \beta^{[l]}$$

$$A^{[l]} = g^{[l]}\left(Z_{BN}^{[l]}\right)$$

where:

- $\mu^{[l]}$ and $\sigma^{[l]}$ are the mean and standard deviation of $Z^{[l]}$
- $Z_{norm}^{[l]}$ is the normalized $Z^{[l]}$
- $\gamma^{[l]}$ and $\beta^{[l]}$ are the learnable scale and shift parameters
- $Z_{BN}^{[l]}$ is the batch-normalized $Z^{[l]}$

This way, the network learns the optimal mean and variance for each layer's inputs during training. It is important to highlight that this adjustment is performed for each mini batch of m examples during every training iteration, which is why the term is known as "batch normalization." During the training phase, the mean (mu) and standard deviation (sigma) are computed for each mini-batch. However, during the prediction phase, we utilize global (population) statistic of mu and sigma. This is obtained by maintaining an exponentially weighted moving average of the mini-batch means and variances calculated during the training phase.

The main purpose of batch normalization is to speed up the convergence and make the network more stable during training by normalizing the inputs. However, it also has a positive side effect related to covariate shift (Shimodaira, 2000). Covariate shift refers to the change in the input distribution to a learning system, which can lead to performance degradation. Batch normalization helps mitigate this by normalizing the inputs. Additionally, batch normalization has a regularization effect (Luo et al., 2018), similar to dropout, as the statistics are computed on different mini-batches with different instances, adding some noise to the model's internal representations. This noise can have a slight regularizing effect.

The Python implementation of batch normalization is straightforward and shown in the below code for reference for a three-layer neural network:

Listing 5.8 Forward Propagation with Batch Normalization

```
def batchnorm_forward(Z, gamma, beta):
    mu = np.mean(Z, axis=0)
    var = np.var(Z, axis=0)
    Z_norm = (Z - mu) / np.sqrt(var + 1e-8)
    Z_BN = gamma * Z_norm + beta

    return Z_BN

def forward_propagation_batchnorm(X, parameters):
    W1 = parameters['W1']
    b1 = parameters['b1']
    gamma1 = parameters['gamma1']
    beta1 = parameters['beta1']

    W2 = parameters['W2']
    b2 = parameters['b2']
    gamma2 = parameters['gamma2']
    beta2 = parameters['beta2']

    W3 = parameters['W3']
    b3 = parameters['b3']
    gamma3 = parameters['gamma3']
    beta3 = parameters['beta3']

    1 = np.dot(W1, X) + b1
    Z1_BN = batchnorm_forward(Z1, gamma1, beta1)
    A1 = relu(Z1_BN)

    2 = np.dot(W2, A1) + b2
    Z2_BN = batchnorm_forward(Z2, gamma2, beta2)
    A2 = relu(Z2_BN)
```

(continued)

Listing 5.8 (continued)

```
    Z3 = np.dot(W3, A2) + b3
    Z3_BN = batchnorm_forward(Z3, gamma3, beta3)
    A3 = Z3_BN

    return A3
```

5.6 Summary

- Batch size variations in gradient descent include full batch, mini batch, and stochastic gradient descent, with each having practical applications based on dataset size.
- Momentum-based gradient descent overcomes local minima and improves convergence speed, addressing limitations of basic gradient descent.
- Momentum-based gradient descent uses velocity terms for parameter updates.
- The Adam optimization algorithm combines momentum and RMSprop concepts, offering efficient training.
- Learning rate decay balances training speed and convergence precision.
- Adam optimization shows faster convergence compared to traditional gradient descent.

References

Bengio, Y. (2012). Practical recommendations for gradient-based training of deep architectures. In *Neural networks: Tricks of the trade: Second edition* (pp. 437–478). Springer.

Bottou, L. (2010). Large-scale machine learning with stochastic gradient descent. *Proceedings of COMPSTAT'2010: 19th International Conference on Computational StatisticsParis France, August 22-27, 2010 Keynote, Invited and Contributed Papers*, 177–186.

Dauphin, Y. N., Pascanu, R., Gulcehre, C., Cho, K., Ganguli, S., & Bengio, Y. (2014). Identifying and attacking the saddle point problem in high-dimensional non-convex optimization. *Advances in Neural Information Processing Systems, 27.*

Ioffe, S., & Szegedy, C. (2015). Batch normalization: Accelerating deep network training by reducing internal covariate shift. *International Conference on Machine Learning*, 448–456.

Kingma, D. P. (2014). Adam: A method for stochastic optimization. *arXiv Preprint arXiv:1412.6980.*

Li, M., Zhang, T., Chen, Y., & Smola, A. J. (2014). Efficient mini-batch training for stochastic optimization. *Proceedings of the 20th ACM SIGKDD International Conference on Knowledge Discovery and Data Mining*, 661–670.

Luo, P., Wang, X., Shao, W., & Peng, Z. (2018). Towards understanding regularization in batch normalization. *arXiv Preprint arXiv:1809.00846.*

Osborne, J. (2010). Improving your data transformations: Applying the Box-Cox transformation. *Practical Assessment, Research, and Evaluation, 15*(1).

Polyak, B. T. (1964). Some methods of speeding up the convergence of iteration methods. *Ussr Computational Mathematics and Mathematical Physics*, *4*(5), 1–17.

Qian, N. (1999). On the momentum term in gradient descent learning algorithms. *Neural Networks*, *12*(1), 145–151.

Santurkar, S., Tsipras, D., Ilyas, A., & Madry, A. (2018). How does batch normalization help optimization? *Advances in Neural Information Processing Systems*, *31*.

Seber, G. A., & Lee, A. J. (2003). *Linear regression analysis*. John Wiley & Sons.

Shimodaira, H. (2000). Improving predictive inference under covariate shift by weighting the log-likelihood function. *Journal of Statistical Planning and Inference*, *90*(2), 227–244.

Sutskever, I., Martens, J., Dahl, G., & Hinton, G. (2013). On the importance of initialization and momentum in deep learning. *International Conference on Machine Learning*, 1139–1147.

Tieleman, T., & Hinton, G. (2012). Lecture 6.5-rmsprop, coursera: Neural networks for machine learning. *University of Toronto, Technical Report*, *6*.

Willmott, C. J., & Matsuura, K. (2005). Advantages of the mean absolute error (MAE) over the root mean square error (RMSE) in assessing average model performance. *Climate Research*, *30*(1), 79–82.

Xu, D., Zhang, S., Zhang, H., & Mandic, D. P. (2021). Convergence of the RMSProp deep learning method with penalty for nonconvex optimization. *Neural Networks*, *139*, 17–23.

You, K., Long, M., Wang, J., & Jordan, M. I. (2019). How does learning rate decay help modern neural networks? *arXiv Preprint arXiv:1908.01878*.

Zeiler, M. D. (2012). Adadelta: An adaptive learning rate method. *arXiv Preprint arXiv:1212.5701*.

Chapter 6
Applying Convolutional Neural Networks

This chapter covers:

- Introducing convolution operations with padding and striding.
- Using pooling layer in convolutional neural networks (CNNs).
- Understanding CNN parameters and output dimension.
- Coding initialization and training process of CNN parameters.
- Demonstrating a case study: Citrus leaf disease classification using CNNs.

Convolutional neural networks (CNNs) are one of the key architectures available to build deep learning models. These networks have revolutionized image processing tasks by leveraging their ability to automatically learn hierarchical representations directly from pixel data. This chapter unfolds the fundamental principles of CNNs, from convolution operations to the practical implementation of an end-to-end image classification model. We will look into the convolutional and pooling layers that define their structure, examine how parameters are initialized for effective learning, and discuss forward propagation to understand how CNNs transform input data into meaningful predictions. Additionally, we will apply these concepts to a real-world example: the classification of citrus leaf diseases using an end-to-end CNN model. We will gather knowledge and skills to use CNNs for image processing tasks.

6.1 Neural Networks in Computer Vision

Computer vision tasks primarily involve the identification and interpretation of objects within images and videos. The videos are essentially sequences of individual frames or images. So, how can we effectively apply neural networks to image datasets? One naive approach could be to utilize fully connected neural networks (Goodfellow 2016), as we have used for tabular data in the prior chapters. However, utilizing the fully connected neural networks to images presents certain challenges.

T. Islam, *Hands-on Deep Learning*, https://doi.org/10.1007/978-3-032-00488-8_6

Let us look into the limitations of fully connected neural networks for images and videos:

- **High dimensionality**: Images are inherently high-dimensional data. Let us take an example of a resized RGB image with a dimension of 100 × 100 × 3. This will produce 30,000 features just for an example image. This high-dimensionality produces significant challenges in applying fully connected neural networks to images.
- **High parameter count**: In a fully connected network, every node in each layer is connected to every node in the next layer. For a 100 × 100 pixel input image, a single hidden layer with just 1000 nodes would require 100 × 100 × 1000 = 10,000,000 weights. This high number of parameters makes the network computationally expensive to train.
- **Overfitting**: The high number of parameters also makes the network prone to overfitting, especially when the amount of training data is limited.
- **Lack of spatial invariance**: The spatial structure of the image is not preserved during learning in the fully connected neural networks.

Convolutional neural networks (CNNs) address these issues by introducing the concept of convolutional layers and pooling layers (Gu et al. 2018; LeCun et al. 2002). These layers have automated feature extraction capability (Zeiler and Fergus 2014). This means important features such as edges, corners, and color blobs can be detected automatically, eliminating the need of extracting features manually from the image.

In CNN architecture, the convolution is "locally connected." This means that each neuron in a convolutional layer is connected only to a small, localized region of the input data, rather than to every neuron in the previous layer. In contrast, a fully connected layer links each neuron to every neuron in the preceding layer, making it computationally expensive. For tasks such as image processing, local patterns are more important, so convolution is used. However, a fully connected layer is generally used in the final layer (Krizhevsky et al. 2012).

A typical CNN architecture consists of several layers:

1. *Convolutional layer*: This layer performs convolution operations and applies a set of filters or kernels to the input image. These filters are essentially weights that are learnable.
2. *Pooling layer*: This layer reduces the spatial size of the representation, thereby decreasing the number of parameters in the network, thus controlling overfitting.
3. *Fully connected layer*: After several convolutional and pooling layers, fully connected layers are added. This is done after flattening the previous layer.

We are already familiar with the fully connected layer, as discussed in prior chapters. Now, let us look into the concepts of the convolutional layer and pooling layer.

6.2 Convolutional Layer

6.2.1 Convolution Operation

The convolution operation in a convolutional layer involves overlaying a kernel (also known as a filter) onto the input matrix and performing element-wise multiplication between the kernel and the corresponding elements of the input matrix (Dumoulin and Visin 2016). The results of these multiplications are then summed to produce a single value in the output matrix. This process is repeated across the entire spatial extent of the input matrix, with the kernel sliding (or convolving) over the input. An example of convolution operation is illustrated in Fig. 6.1 by using a vertical edge detection filter.

The Python implementation of convolution operation will look like this:

Listing 6.1 Convolution Operation

```
import numpy as np

def convolution_operation(input_matrix, kernel):
    # Dimensions of the input matrices
    w, h = input_matrix.shape
    kw, kh = kernel.shape

    # Dimensions of the output matrix
    output_w = w - kw + 1
    output_h = h - kh + 1

    # Initialize the output matrix
    output_matrix = np.zeros((output_w, output_h))

    # Apply the kernel to each position of the input matrix
    for i in range(output_w):
        for j in range(output_h):
            output_matrix[i][j] = np.sum(input_matrix[i:i+kw,
j:j+kh] * kernel)
    return output_matrix
```

Now, we can plug-in the input matrix to the above function to produce the output matrix from convolution operation. Let us plug-in our example onto the above function:

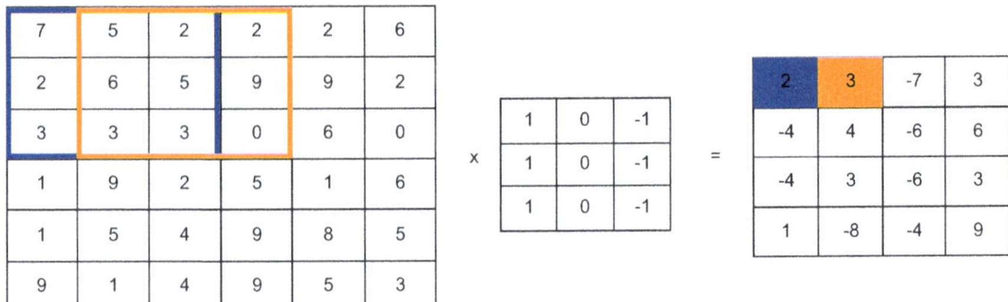

Fig. 6.1 Illustration of convolution operation

```
# Define a 5x5 input matrix
input_matrix = np.array(
    [[7, 5, 2, 2, 2, 6],
     [2, 6, 5, 9, 9, 2],
     [3, 3, 3, 0, 6, 0],
     [1, 9, 2, 5, 1, 6],
     [1, 5, 4, 9, 8, 5],
     [9, 1, 4, 9, 5, 3]])
# Define a 3x3 filter/kernel
filter = np.array([[1, 0, -1],
                   [1, 0, -1],
                   [1, 0, -1]])
# Apply the convolution operation
output_matrix = convolution_operation(input_matrix, filter)
print(output_matrix)
```

```
[[ 2. 3. -7. 3.]
 [-4. 4. -6. 6.]
 [-4. 3. -6. 3.]
 [ 1. -8. -4. 9.]]
```

For the sake of understanding how the convolution operation works, let us take the first element of the output matrix as an example, which is 2. The first 3×3 block of the input matrix is:

```
7 5 2
2 6 5
3 3 3
```

And the kernel or filter is:

```
1 0 -1
1 0 -1
1 0 -1
```

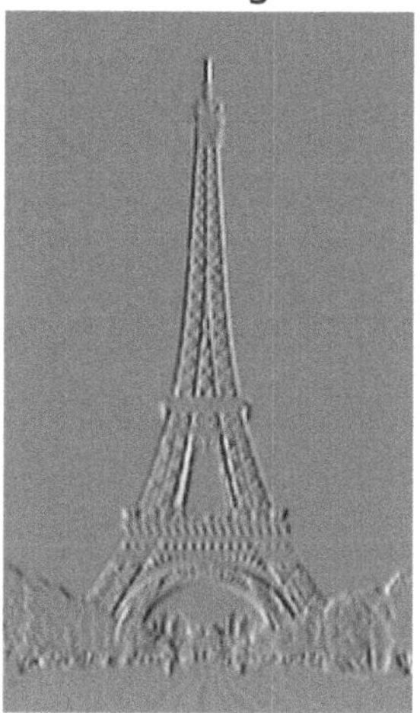

Fig. 6.2 Illustration of vertical and horizontal edge detection

The convolution operation for the first element is calculated as follows:

```
(7*1 + 5*0 + 2*-1) +
(2*1 + 6*0 + 5*-1) +
(3*1 + 3*0 + 3*-1) = 2
```

This gives us the first element of the output matrix, which is 2. Similarly, the second element of the output matrix is 3, and so forth.

Now that we have an understanding of how the basic convolution operation works, one might wonder about the nature of the filters used and how they are determined. Over the years, computer vision researchers have developed various types of filters tailored for different tasks such as edge detection (Marr and Hildreth 1980). These filters are designed to highlight specific features in images, such as vertical or horizontal edges. For instance, consider the examples of vertical and horizontal edge detection demonstrated in Fig. 6.2 by applying two distinct filters. The vertical edge detection filter used in this image is:

```
[1, 0, -1],
[1, 0, -1],
[1, 0, -1]
```

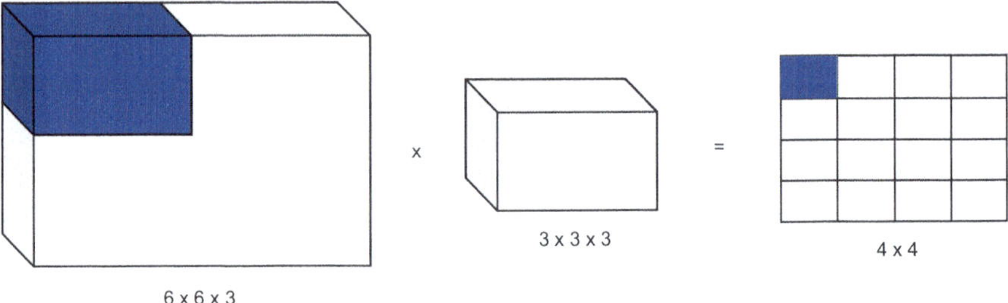

Fig. 6.3 Illustration of convolutions over volumes

In contrast, the horizontal edge detection filter applied in this image is:

```
[1, 1, 1],
[0, 0, 0],
1;[-1, -1, -1]
```

As we examine the figure, we can clearly see the outcomes of applying these filters. The vertical edges of the Eiffel Tower structure are emphasized by the vertical edge detection filter, while the horizontal edges are accentuated by the horizontal edge detection filter. The vertical and horizontal edges are just a few examples of specific features that can be extracted from images using convolutional filters. In reality, images contain a multitude of distinct features beyond simple edges. These features can range from textures, patterns, and shapes to more complex structures and objects. As we can imagine, developing filters manually for specific features in images can indeed be a challenging and labor-intensive process. This is where the power of convolutional neural networks (CNNs) becomes apparent. CNNs are designed to automatically learn and optimize these filters, alleviating the need for manual filter engineering. Each filter specializes in recognizing specific patterns or structures within the input image data. These filters are essentially the parameters or weights learned through backpropagation and gradient descent.

So far, we have explored the convolution operation in the context of 2D grayscale images. However, most images are represented in RGB format, containing three color channels: red, green, and blue. Therefore, when applying convolution in the context of RGB images, we will have to extend the operation to a volume rather than a 2D matrix. Convolutions over volumes involve performing the same operation as discussed earlier, but across the RGB dimension. This is depicted in Fig. 6.3. Consider an example where we have a $6 \times 6 \times 3$ RGB image or input matrix. When applying a convolution operation, the filter dimensions become $3 \times 3 \times 3$, representing a volume operation across all three RGB channels. As a result, the dimension of the final output matrix will be 4×4. Each element in this 4×4 matrix is computed by summing the element-wise products of the selected $3 \times 3 \times 3$ region of

0	0	0	0	0	0	0	0
0	7	5	2	2	2	6	0
0	2	6	5	9	9	2	0
0	3	3	3	0	6	0	0
0	1	9	2	5	1	6	0
0	1	5	4	9	8	5	0
0	9	1	4	9	5	3	0
0	0	0	0	0	0	0	0

x

1	0	-1
1	0	-1
1	0	-1

=

-11	2	0	-4	3	11
-14	2	3	-7	3	17
-18	-4	4	-6	6	16
-17	-4	4	-6	3	15
-15	1	-8	-4	9	14
-6	2	-12	-5	10	13

Fig. 6.4 Illustration of convolution operation with "same" padding

the input volume and the corresponding 3 × 3 × 3 kernel filter volume. For instance, the value at the top-left corner of the 4 × 4 output matrix is derived from the sum of all 27 element-wise multiplications between the selected 3 × 3 × 3 region of the input RGB volume and the 3 × 3 × 3 kernel volume. This process is repeated across the entire input volume to produce the final convolutional output, capturing spatial patterns and features across all RGB color channels.

This means convolution operations over RGB images involve processing volumetric data and the convolutional filters operate across these three dimensions. However, it is important to note that image data can have more than just three channels, especially in specialized contexts such as NASA's hyper-spectral images. These hyper-spectral images can contain hundreds of channels, each representing a different wavelength or spectral band (Yokoya et al. 2017). In such cases, the same volumetric convolution operation applies, but across a higher-dimensional space defined by the numerous channels present in the image or image-like matrices.

6.2.2 *Padding*

Padding involves adding rows and columns of zeros around an input matrix or image before applying the convolution operation (Chollet 2021; Dumoulin and Visin 2016). The purpose of padding is to ensure that the dimension of the output matrix or image remains the same as that of the input matrix after convolution. Specifically, when padding is applied to preserve the input–output dimensionality, it is known as "same" padding. This ensures that the spatial dimensions of the input and output remain consistent throughout the convolution process. On the other hand, when no padding is used and the convolution operation is applied directly to the input matrix, resulting in a smaller output matrix (due to the reduction in spatial dimensions), it is referred to as "valid" padding. Figure 6.1 is an example of "valid" padding or no padding. We can see that dimension of the output matrix is reduced in this case. On the other hand, the "same" padding retains the original dimensions of the input matrix as shown in Fig. 6.4.

The Python implementation of padding will look like this:

Listing 6.2 Padding Operation

```
def add_padding(input_matrix, padding_type="same", filter_size=3):
    if padding_type == "same":
        input_rows, input_cols = input_matrix.shape
        pad_rows = max(filter_size - 1, 0)
        pad_cols = max(filter_size - 1, 0)

        top_pad = pad_rows // 2
        bottom_pad = pad_rows - top_pad
        left_pad = pad_cols // 2
        right_pad = pad_cols - left_pad

        padded_matrix = np.pad(
            input_matrix,
            ((top_pad, bottom_pad), (left_pad, right_pad)),
            mode="constant",
            constant_values=0,
        )
    elif padding_type == "valid":
        padded_matrix = np.copy(input_matrix)
    return padded_matrix
```

Now, let us look at our example input matrix with the dimension 6 × 6. We want to apply both same and valid padding on the input matrix:

```
input_matrix = np.array(
    [
        [7, 5, 2, 2, 2, 6],
        [2, 6, 5, 9, 9, 2],
        [3, 3, 3, 0, 6, 0],
        [1, 9, 2, 5, 1, 6],
        [1, 5, 4, 9, 8, 5],
        [9, 1, 4, 9, 5, 3],
    ]
)

# Add 'same' padding
padded_matrix_same = add_padding(input_matrix, padding_type="same",
filter_size=3)
```

```
print("Input Matrix:")
print(input_matrix)
print("Padded Matrix with 'same' padding")
print(padded_matrix_same)

# Apply the convolution operation
output_matrix = convolution_operation(padded_matrix_same, filter)
print("Output Matrix with 'same' padding")
print(output_matrix)

# Add 'valid' padding (no padding)
padded_matrix_valid = add_padding(input_matrix,
padding_type="valid", filter_size=3)
print("Padded Matrix with 'valid' padding")
print(padded_matrix_valid)

# Apply the convolution operation
output_matrix = convolution_operation(padded_matrix_valid, filter)
print("Output Matrix with 'valid' padding")
print(output_matrix)
```

This will result in the following outputs:

```
Input Matrix:
[[7 5 2 2 2 6]
 [2 6 5 9 9 2]
 [3 3 3 0 6 0]
 [1 9 2 5 1 6]
 [1 5 4 9 8 5]
 [9 1 4 9 5 3]]
Padded Matrix with 'same' padding
[[0 0 0 0 0 0 0 0]
 [0 7 5 2 2 2 6 0]
 [0 2 6 5 9 9 2 0]
 [0 3 3 3 0 6 0 0]
 [0 1 9 2 5 1 6 0]
 [0 1 5 4 9 8 5 0]
 [0 9 1 4 9 5 3 0]
 [0 0 0 0 0 0 0 0]]
Output Matrix with 'same' padding
[[-11.  2.   0. -4.   3. 11.]
 [-14.  2.   3. -7.   3. 17.]
 [-18. -4.   4. -6.   6. 16.]
 [-17. -4.   3. -6.   3. 15.]
```

```
 [-15.  1.  -8. -4.  9. 14.]
 [ -6.  2. -12. -5. 10. 13.]]
Padded Matrix with 'valid' padding
[[7 5 2 2 2 6]
 [2 6 5 9 9 2]
 [3 3 3 0 6 0]
 [1 9 2 5 1 6]
 [1 5 4 9 8 5]
 [9 1 4 9 5 3]]
Output Matrix with 'valid' padding
[[ 2.  3. -7. 3.]
 [-4.  4. -6. 6.]
 [-4.  3. -6. 3.]
 [ 1. -8. -4. 9.]]
```

We can observe that the convolution operation with 'same' padding has resulted in a 6 × 6 matrix, which retains the original dimensions of the input image. This 'same' padding ensures that the output matrix size matches the input size after convolution. On the other hand, when using 'valid' padding, the dimension of the output matrix is reduced to 4 × 4. 'Valid' padding means no padding is added around the input matrix, resulting in a smaller output size compared to the input size. This reduction in dimension occurs because the convolution operation does not extend beyond the boundaries of the input matrix, resulting in a smaller spatial extent of the output feature map.

6.2.3 Striding

Striding is a concept in convolution operation that controls how the kernel or filter matrix slides around the input matrix. So far, all our examples have been with stride = 1. This means the kernel slides one pixel or step at a time during convolution operation. Let us now take a look at an example (Fig. 6.5), where we illustrate the

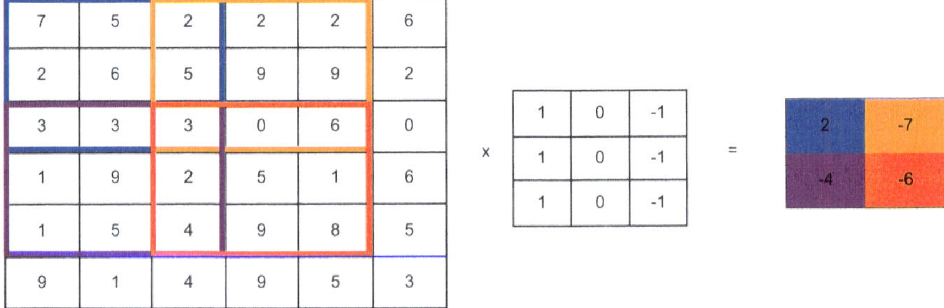

Fig. 6.5 Illustration of striding in convolution operation

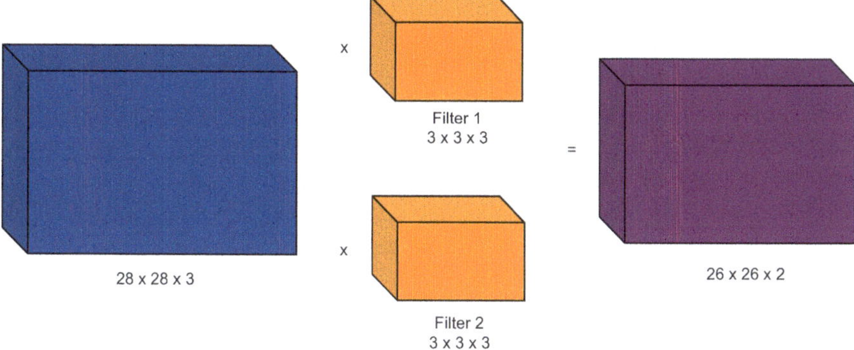

Fig. 6.6 Illustration of Conv2D with multiple filters

impact of using a stride of 2 during convolution with a 3 × 3 filter on a 6 × 6 input matrix, resulting in a 2 × 2 output matrix. The convolution process begins at the top-left corner of the input matrix, where each element within a 3 × 3 region is multiplied with the corresponding element in the filter. Summing these products yields the top-left element of the output matrix. We then move 2 steps to the right (according to the specified stride) to process the next element in the output matrix. This process continues horizontally until the end of the row is reached. Subsequently, we move down 2 steps (as per the stride value) and repeat the process for the next row. The convolution process stops when further downward movement would go beyond the input matrix boundaries.

6.2.4 Bringing It All Together

Now that we have introduced convolution operations with striding and padding, let us integrate these concepts together. One crucial aspect we have not discussed yet is the use of multiple filters and kernels in convolution operations (Simonyan and Zisserman 2014). Up to this point, our examples have focused on understanding the process using a single filter. However, in practical applications, convolutional layers typically involve applying multiple filters and stacking their outputs. The rationale behind using multiple filters is to capture various image properties. For example, different filters can specialize in detecting different properties, such as vertical edges, horizontal edges, textures, or other visual features. Figure 6.6 provides a visual example of this concept. Consider an input image of size 28 × 28 × 3. If we apply two 3 × 3 kernels to this image for convolution without padding, the resulting output matrix will be 26 × 26 × 2. This outcome reflects the stacking of two kernels in the convolution operation, with each kernel contributing to a distinct feature representation within the output matrix.

Let us now code up a complete convolutional layer from scratch bringing it all together:

Listing 6.3 Convolutional Layer

```
def conv2d(X, W, stride=1, padding=0):

    batch_size, input_height, input_width, in_channels = X.shape
    filter_height, filter_width, _, out_channels = W.shape

    # Calculate output dimensions
    output_height = (input_height - filter_height + 2 * padding) //
stride + 1
    output_width = (input_width - filter_width + 2 * padding) // stride + 1

    # Apply padding to the input
    if padding > 0:
        X_padded = np.pad(
            X, ((0, 0), (padding, padding), (padding, padding),
(0, 0)), mode="constant"
        )
    else:
        X_padded = X

    # Initialize output tensor
    Z = np.zeros((batch_size, output_height, output_width,
out_channels))

    # Perform convolution
    for b in range(batch_size):
        for h_out in range(output_height):
        for w_out in range(output_width):
            h_start = h_out * stride
            h_end = h_start + filter_height
            w_start = w_out * stride
            w_end = w_start + filter_width

            # Extract the receptive field from the input
            X_slice = X_padded[b, h_start:h_end, w_start:w_end, :]

            # Apply filter and sum over spatial dimensions
            for c_out in range(out_channels):
                filter_slice = W[:, :, :, c_out]
                ;Z[b, h_out, w_out, c_out] = np.sum(X_slice *
filter_slice)

    return Z
```

The conv2d function performs a 2D convolution operation on an input tensor X using a convolutional filter W. The stride and padding parameters are used to control extent of striding and padding to be applied. The function calculates the output dimensions based on the input size, filter size, padding, and stride settings. During convolution, the function applies the filter to extract features from the input tensor, producing an output tensor with specified dimensions and channel depth determined by the number of filters in W. The resulting output represents the feature maps generated by convolving W across X with specified stride and padding configurations.

We can validate the accuracy of our custom Conv2D implementation by comparing its output with that of TensorFlow's Conv2D function. Given the same input matrix 'X' and filter matrix 'W', both implementations should yield identical results for the 'Z' calculation:

```
X = np.random.rand(1, 28, 28, 3)
W = np.random.rand(3, 3, 3, 2)
Z_np = conv2d(X, W, stride=1, padding=0)
Z_tf = tf.nn.conv2d(X, W, strides=[1, 1, 1, 1], padding='VALID')
print (np.allclose(Z_np, Z_tf))
```

```
True
```

Please note that the aforementioned convolution operation does not include the bias term. Therefore, it is necessary to add this term separately. Once the value of Z is computed, we can apply a nonlinear activation function to calculate the activation A. In summary, a convolutional layer comprises these two key computations:

$$Z = Conv2D(W) + b$$

$$A = Relu(Z)$$

6.3 Pooling Layer

The pooling layer is a fundamental component in the architecture of convolutional neural networks (CNNs). Its primary function is to progressively reduce the spatial size of the feature map representation, thereby decreasing the number of parameters and computational complexity in the network. This layer performs a down-sampling operation along the spatial dimensions, resulting in a condensed representation of the input feature maps (Scherer et al. 2010). There are two main types of pooling operations: Maximum Pooling and Average Pooling.

Maximum Pooling, also known as Max Pooling, operates by selecting the maximum value from each segment of the input matrix within a specified filter

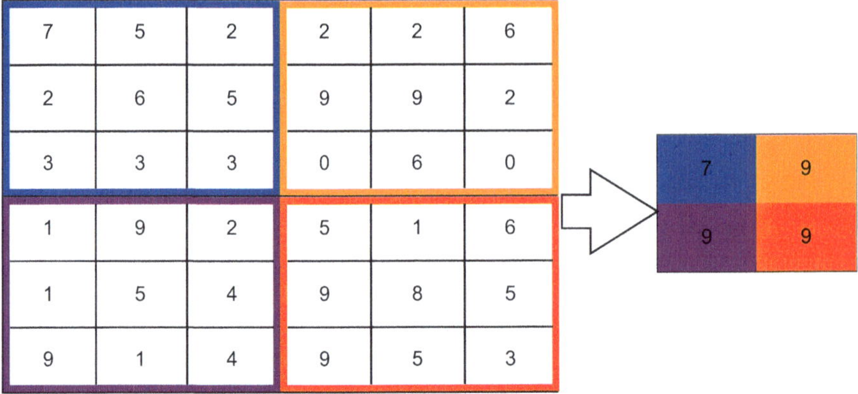

Fig. 6.7 Illustration of maximum pooling operation

size. This has been a popular choice among computer vision researchers due to its effectiveness in capturing the most salient features in the input.

Average Pooling, on the other hand, calculates the average of all values within the filter, providing a smoothed representation of the input.

Similar to the convolution operation, the key hyperparameters for the pooling layer are the filter size and the stride. The filter size defines the extent of the down-sampling, while the stride determines the step size when moving the filter across the input matrix. Let us consider an example of a maximum pooling operation with a 3×3 filter size and a stride of 3 (Fig. 6.7). The value 7 is the maximum value in the top-left 3×3 segment of the input matrix. The value 9 is the maximum value in the top-right 3×3 segment. Similarly, the bottom row of the output matrix contains the maximum values from the corresponding 3×3 segments in the bottom half of the input matrix. It is worth noting that there are no parameters to learn. As such, we can see that there is no filter matrix involved in the pooling operation.

Let us now implement maximum pooling operation from scratch using numpy:

Listing 6.4 Maximum Pooling Operation

```
def maxpool_2d(X, pool_size=2, stride=2, padding=0):

    # Determine the pooling window size (pool_height, pool_width)
    if isinstance(pool_size, int):
        pool_height = pool_size
        pool_width = pool_size
    else:
        pool_height, pool_width = pool_size

    # Determine the stride size (stride_height, stride_width)
    if isinstance(stride, int):
```

(continued)

Listing 6.4 (continued)

```
        stride_height = stride
        stride_width = stride
    else:
        stride_height, stride_width = stride

    # Get input tensor dimensions
    batch_size, input_height, input_width, channels = X.shape

    # Calculate output dimensions
    output_height = (input_height - pool_height + 2 * padding) //
stride_height + 1
    output_width = (input_width - pool_width + 2 * padding) //
stride_width + 1

    # Apply padding to the input
    if padding > 0:
        X_padded = np.pad(X, ((0, 0), (padding, padding), (padding,
padding), (0, 0)), mode='constant')
    else:
        X_padded = X

    # Initialize output tensor
    pooled_output = np.zeros((batch_size, output_height,
output_width, channels))

    # Perform max pooling using array slicing and aggregation
    for b in range(batch_size):
        for h_out in range(output_height):
            for w_out in range(output_width):
                h_start = h_out * stride_height
                h_end = h_start + pool_height
                w_start = w_out * stride_width
                w_end = w_start + pool_width

                # Extract the receptive field from the padded input
                X_slice = X_padded[b, h_start:h_end, w_start:w_end, :]

                # Perform max pooling over spatial dimensions
                pooled_output[b, h_out, w_out, :] = np.amax
(X_slice, axis=(0, 1))

    return pooled_output
```

We can validate our maxpool_2d implementation against TensorFlow's MaxPool2D function. Given the same input matrix X, both implementations should yield identical results for the pooled output calculation:

```
X = np.random.rand(1, 28, 28, 3)
pooled_output_np = maxpool_2d(X, pool_size=2, stride=2, padding=0)
pooled_output_tf = tf.nn.max_pool(X, ksize=[1, 2, 2, 1], strides=[1,
2, 2, 1], padding='VALID')
print (np.allclose(pooled_output_np, pooled_output_tf))
```

```
True
```

6.4 Output Dimension

It is important that, we get our output dimension right in building a CNN. There is a formula that can be used for both convolutional and pooling operation. Let us assume, we have an input matrix with the dimension $(n_{H_{in}}, n_{W_{in}}, n_c)$. After applying convolutional or pooling operation, the dimension of the output matrix then will be $(n_{H_{out}}, n_{W_{out}}, n_c)$ based on the following formulas:

$$n_{H_{out}} = \left\lfloor \frac{n_{H_{in}} - f + 2p}{s} \right\rfloor + 1$$

$$n_{W_{out}} = \left\lfloor \frac{n_{W_{in}} - f + 2p}{s} \right\rfloor + 1$$

$$n_C = \text{number of filters (conv) or number of channels (pool)}$$

where:

- $n_{H_{in}}$ is the height of the previous layer,
- $n_{H_{out}}$ is the width of the previous layer,
- f is the size of the filter,
- p is the padding,
- s is the stride.

This formula ensures that we have an integer output height. If the result of the division is not an integer, it is rounded down to the nearest integer. Let us now implement this:

Listing 6.5 Calculating Output Dimensions

```
def calculate_output_dims(nHin, nWin, nCin, f, p, s, operation=
'conv', nFilters=None):

    nHout = int((nHin - f + 2*p) // s) + 1
    nWout = int((nWin - f + 2*p) // s) + 1

    if operation == 'conv':
        nCout = nFilters
    elif operation == 'pool':
        nCout = nCin

    return (nHout, nWout, nCout)
```

Now that we have implemented the function to compute the output dimensions, let us look into an illustrative example. We will apply our function to sequentially calculate the output shape of the first block in the VGG16 architecture (Simonyan and Zisserman 2014). The first block of the VGG16 architecture comprises two convolution operations, followed by a max pooling operation. The convolutional layers, each with a 3×3 filter size, stride of 1, and padding of 1, transform an input feature map of size $224 \times 224 \times 3$ into an output shape of $224 \times 224 \times 64$. The subsequent max pooling layer, with a 2×2 filter size and stride of 2, further transforms the feature map into an output shape of $112 \times 112 \times 64$. This can be confirmed by our `calculate_output_dims` function as shown below:

```
nHout, nWout, nCout = calculate_output_dims(nHin=224, nWin=224,
nCin=3, f=3, p=1, s=1, operation='conv', nFilters=64)
print ("block1_conv1 (Conv2D) output shape:", nHout, nWout, nCout)
nHout, nWout, nCout = calculate_output_dims(nHin=224, nWin=224,
nCin=64, f=3, p=1, s=1, operation='conv', nFilters=64)
print ("block1_conv2 (Conv2D) output shape:", nHout, nWout, nCout)
nHout, nWout, nCout = calculate_output_dims(nHin=224, nWin=224,
nCin=64, f=2, p=0, s=2, operation='pool')
print ("block1_pool (Maxpool2D) output shape:", nHout, nWout, nCout)
```

```
block1_conv1 (Conv2D) output shape: 224 224 64
block1_conv2 (Conv2D) output shape: 224 224 64
block1_pool (Maxpool2D) output shape: 112 112 64
```

6.5 Flattening

Flattening is a concept of the transformation of the high-dimensional output after convolutional and pooling layers operation into a long continuous linear vector. This is because fully connected layers are often used in a CNN architecture following convolutional and pooling layers. The fully connected layer requires a 1D input vector. Therefore, the flattening is a connection between the convolution and/or pooling output to the fully connected layer. In Fig. 6.8, we visually illustrate the process of the flattening operation. This operation transforms an input matrix with dimensions of $7 \times 7 \times 512$ into a vector with a dimension of 25,088. The transformation effectively "flattens" the three-dimensional matrix into a one-dimensional vector, hence the term "flattening." This operation is crucial in preparing the data for subsequent layers in the network.

6.6 Building an End-to-End CNN Model

We should now have a fundamental understanding of the core operations of a convolutional neural network (CNN). This means we are now prepared to build a comprehensive architecture. Let us now build an end-to-end CNN model for a practical application.

6.6.1 *Image Classification Problem*

We will now build an end-to-end CNN model for image classification, utilizing the citrus leaves dataset (Rauf et al. 2019). This dataset is a compilation of images categorized into four distinct classes of plant diseases: Black Spot, Canker,

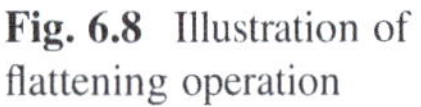

Fig. 6.8 Illustration of flattening operation

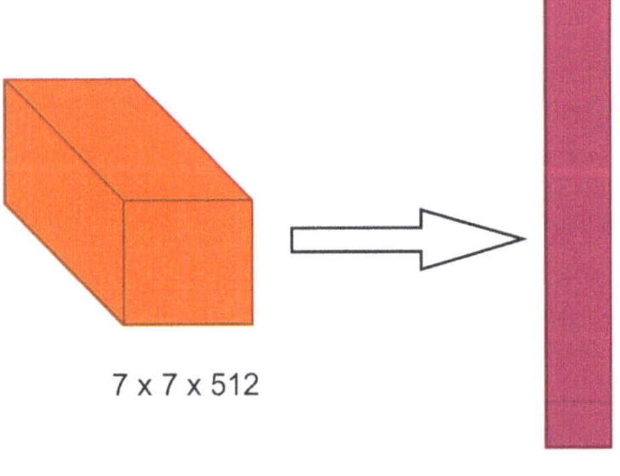

Fig. 6.9 Example images of the citrus leaves dataset

Greening, and Healthy. Figure 6.9 provides few example images of the dataset. Our task is to develop an image classification model to identify whether the citrus plant is healthy or diseased. Plant diseases significantly contribute to agricultural production decline, leading to economic losses. Globally, citrus plants are a key provider of essential nutrients such as vitamin C. Therefore, the classification system has the potential to enhance production by enabling early detection and treatment of diseases.

As usual, we will start by creating a function to load the data:

Listing 6.6 Loading Citrus Leaves Data

```
import tensorflow_datasets as tfds

def load_citrus_leaves_data():
    dataset, info = tfds.load('citrus_leaves', split='train',
shuffle_files=True, with_info=True, as_supervised=True)

    return dataset
```

Let us load the data now:

```
dataset = load_citrus_leaves_data()
```

Next, the images in this dataset, approximately 256 × 25 pixels in resolution, will be resized to a standard 28 × 28 × 3 dimensional RGB format. Besides, we will also have to normalize the image values. The following lines of code show how to apply this preprocessing step:

Listing 6.7 Preprocessing Image

```
def preprocess(image, label):
    image = tf.image.resize(image, [28, 28]) # Resize to 128x128
    image = tf.cast(image, tf.float32) / 255.0 # Normalize pixel values
    return image, tf.one_hot(label, depth=4) # One-hot encode labels
```

We can then create X and y:

```
dataset = dataset.map(preprocess)
X, y = zip(*tfds.as_numpy(dataset))
```

Next, we split the preprocessed dataset into train and test set:

Listing 6.8 Function to Split Dataset into Train and Test Set

```
from sklearn.model_selection import train_test_split

def split_dataset(X, y, test_size=0.2, random_state=42):
    X_train, X_test, y_train, y_test = train_test_split(X, y,
test_size=test_size, random_state=random_state)

    X_train = tf.convert_to_tensor(X_train)
    y_train = tf.convert_to_tensor(y_train)
    X_test = tf.convert_to_tensor(X_test)
    y_test = tf.convert_to_tensor(y_test)
    return X_train, X_test, y_train, y_test
```

We will then run the `split_dataset` function:

```
X_train, X_test, y_train, y_test = split_dataset(X, y)
```

6.6.2 CNN Representation and Architecture

Our basic CNN architecture will comprise a sequence of layers: a convolutional layer, a pooling layer, another convolutional layer, another pooling layer, a fully connected layer, and finally, a softmax layer for classification. This architecture is visually represented in Fig. 6.10. The input matrix is denoted as *X* or *A0*. We then apply a convolution operation using the filter matrix *W[1]* in the first layer. Following this, we add bias term and compute *Z[1]*. The activation matrix is then calculated through a ReLU operation. The parameters to learn in this layer are *W[1]* and *b[1]*. Next, we carry out the max pooling operation. It is worth noting that there are no parameters to learn in the pooling operation. A combination of a convolution operation followed by a max pooling operation is generally referred to as a "block." In a similar manner, we apply another block of convolution and max pooling operations. We then flatten the activation matrix to feed into a fully connected layer. Our final layer is the output layer, which is comprised of a *Z* calculation followed by a softmax activation operation. This completes our end-to-end CNN architecture.

Now, let us look into the parameters or weights in detail to be trained. We initiate the process with an input matrix. Our input dimension is a $28 \times 28 \times 3$ matrix. It is crucial to note that images can come in a variety of dimensions and sizes. Prior to inputting them into a CNN, it is necessary to resize the images to a fixed dimension. For our image classification problem, the images are resized to a $28 \times 28 \times 3$ dimensional matrix.

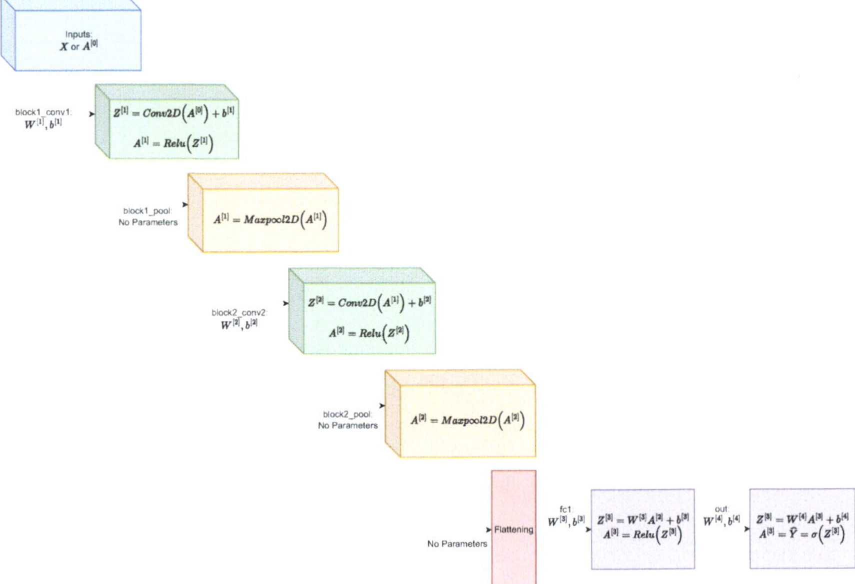

Fig. 6.10 Illustration of a comprehensive end-to-end convolutional neural network (CNN) architecture

TensorFlow provides a nice summary of output shape and parameters. We can cross-check our understanding with TensorFlow output using the following code snippet:

```
model = tf.keras.Sequential([
    tf.keras.layers.Conv2D(32, (3, 3), activation='relu',
input_shape=(28, 28, 3)),
    tf.keras.layers.MaxPooling2D((2, 2)),
    tf.keras.layers.Conv2D(64, (3, 3), activation='relu'),
    tf.keras.layers.MaxPooling2D((2, 2)),
    tf.keras.layers.Flatten(),
    tf.keras.layers.Dense(64, activation='relu'),
    tf.keras.layers.Dense(101, activation='softmax')
])
model.summary()
```

The above code snippet will result in the following output showing output shape and parameters for each layer:

```
Layer (type)                  Output Shape              Param #
=================================================================
conv2d (Conv2D)               (None, 26, 26, 32)        896

max_pooling2d (MaxPooling2   (None, 13, 13, 32)        0
D)

conv2d_1 (Conv2D)             (None, 11, 11, 64)        18496

max_pooling2d_1               (MaxPoolin (None, 5, 5, 64)        0
g2D)

flatten (Flatten)             (None, 1600)              0

dense (Dense)                 (None, 64)                102464

dense_1 (Dense)               (None, 101)               6565

=================================================================
Total params: 128421 (501.64 KB)
Trainable params: 128421 (501.64 KB)
Non-trainable params: 0 (0.00 Byte)
```

Now, let us examine the layers and their parameters:

1. *Conv2D layer*: This is the first convolutional layer of our CNN. The output shape is (None, 26, 26, 32), indicating that the output is a 26 × 26 feature map with 32 filters. The parameter count is 896. These parameters are the weights and biases of the 32 filters used in this layer. Each filter is of size 3 × 3 (from the input depth of 3), so the total number of parameters is 32 × (3 × 3 × 3 + 1) = 896, where the "+1" accounts for the bias term for each filter. A related formula to calculate number of parameters is: $n \times (k1 \times k2 \times c + 1)$, where n = number of filters.
2. *MaxPooling2D layer*: This layer performs a max pooling operation, reducing the spatial dimensions of the input (from 26 × 26 to 13 × 13). This layer has 0 parameters as max pooling does not have any learnable parameters.
3. *Conv2D_1 layer*: This is the second convolutional layer, with an output shape of (None, 11, 11, 64) and 18,496 parameters. The calculation of parameters is similar to the first Conv2D layer, but now we have 64 filters and the input depth is 32 (from the previous layer), so the total parameters are 64 × (3 × 3 × 32 + 1) = 18,496.
4. *MaxPooling2D_1 layer*: Another max pooling layer with 0 parameters, reducing the spatial dimensions from 11 × 11 to 5 × 5.

5. *Flatten layer*: This layer transforms the 3D output from the preceding layer into a 1D vector. It does not have any parameters.
6. *Dense layer*: This is a fully connected layer with an output shape of (None, 64) and 102,464 parameters. The parameters are calculated as (input_size × output_size + output_size) = (1600 × 64 + 64) = 102,464.
7. *Dense_1 layer*: This is the final output layer, using a softmax activation function for multi-class classification. It has an output shape of (None, 101) and 6565 parameters: (input_size × output_size + output_size) = (64 × 101 + 101) = 6565.

In total, this CNN architecture has 128,421 parameters to be trained. These parameters are learned during the training process using backpropagation and gradient descent optimization algorithms. The goal of the training process is to adjust these parameters to minimize the difference between the network's predictions and the actual labels, thereby improving the network's performance on unseen data.

6.6.3 CNN Forward Propagation

The training process of convolutional neural networks (CNNs) begins with the initialization of parameters, which are essentially the weights and biases for each layer in the network. These parameters are typically initialized with small random values, a process that helps break symmetry and allows each neuron to learn something different during training. This is typically done with He initialization (He et al. 2015), a concept we have thoroughly explored in one of the prior chapters.

Consider, for instance, the CNN architecture we previously discussed, which comprises two convolutional blocks (each consisting of a convolution operation followed by pooling). For such an architecture, the parameters can be initialized as follows:

Listing 6.9 Function to Initialize CNN Parameters

```
def initialize_parameters(input_shape, num_classes):
    # Define the model architecture parameters
    filter_size_conv1 = 3
    num_filters_conv1 = 16
    filter_size_conv2 = 3
    num_filters_conv2 = 32
    fc_size = 64

    # Calculate the size of the flattened output after convolutional
and pooling layers
    conv1_output_size = calculate_output_dims(
        nHin=input_shape[0],
```

(continued)

Listing 6.9 (continued)

```
        nWin=input_shape[1],
        nCin=input_shape[2],
        f=filter_size_conv1,
        p=0,
        s=1,
        operation="conv",
        nFilters=num_filters_conv1,
    )
    pool1_output_size = calculate_output_dims(
        nHin=conv1_output_size[0],
        nWin=conv1_output_size[1],
        nCin=conv1_output_size[2],
        f=2,
        p=0,
        s=2,
        operation="pool",
    )
    conv2_output_size = calculate_output_dims(
        nHin=pool1_output_size[0],
        nWin=pool1_output_size[1],
        nCin=pool1_output_size[2],
        f=filter_size_conv2,
        p=0,
        s=1,
        operation="conv",
        nFilters=num_filters_conv2,
    )
    pool2_output_size = calculate_output_dims(
        nHin=conv2_output_size[0],
        nWin=conv2_output_size[1],
        nCin=conv2_output_size[2],
        f=2,
        p=0,
        s=2,
        operation="pool",
    )

    # Initialize parameters for the model
    parameters = {
        # Convolutional Layer 1
        "conv_W1": tf.Variable(
            tf.keras.initializers.HeNormal()(
```

(continued)

Listing 6.9 (continued)

```
                shape=[
                    filter_size_conv1,
                    filter_size_conv1,
                    input_shape[-1],
                    num_filters_conv1,
                ]
            )
        ),
        "conv_b1": tf.Variable(tf.zeros([num_filters_conv1])),
        # Convolutional Layer 2
        "conv_W2": tf.Variable(
            tf.keras.initializers.HeNormal()(
                shape=[
                    filter_size_conv2,
                    filter_size_conv2,
                    num_filters_conv1,
                    num_filters_conv2,
                ]
            )
        ),
        "conv_b2": tf.Variable(tf.zeros([num_filters_conv2])),
        # Fully Connected Layer
        "fc_W": tf.Variable(
            tf.keras.initializers.HeNormal()(
                shape=[
                    pool2_output_size[0] * pool2_output_size[1] *
pool2_output_size[2],
                    fc_size,
                ]
            )
        ),
        "fc_b": tf.Variable(tf.zeros([fc_size])),
        # Output Layer
        "out_W": tf.Variable(
            tf.keras.initializers.HeNormal()(shape=[fc_size,
num_classes])
        ),
        "out_b": tf.Variable(tf.zeros([num_classes])),
    }

    return parameters
```

With the parameters and inputs at hand, we can then proceed to implement the forward propagation as follows. Here, the code defines the forward propagation function with two convolutional layers, a fully connected layer, and an output layer. The first convolutional layer applies a convolution operation followed by a ReLU activation (Nair and Hinton 2010) and max pooling. The second convolutional layer repeats this process. The output from the second convolutional layer is then flattened and passed through a fully connected layer with a ReLU activation. Finally, the output layer applies a softmax activation to produce the final output probabilities. The parameters for the layers, such as weights and biases, are retrieved from the parameters dictionary:

Listing 6.10 Forward Propagation of a CNN

```
def forward_propagation(X, parameters):
    # Retrieve parameters
    conv_W1 = parameters['conv_W1']
    conv_b1 = parameters['conv_b1']
    conv_W2 = parameters['conv_W2']
    conv_b2 = parameters['conv_b2']
    fc_W = parameters['fc_W']
    fc_b = parameters['fc_b']
    out_W = parameters['out_W']
    out_b = parameters['out_b']


    # Convolutional Layer 1
    conv1 = tf.nn.conv2d(X, conv_W1, strides=[1, 1, 1, 1],
padding='VALID') + conv_b1
    conv1_relu = tf.nn.relu(conv1)
    pool1 = tf.nn.max_pool(conv1_relu, ksize=[1, 2, 2, 1],
strides=[1, 2, 2, 1], padding='VALID')


    # Convolutional Layer 2
    conv2 = tf.nn.conv2d(pool1, conv_W2, strides=[1, 1, 1, 1],
padding='VALID') + conv_b2
    conv2_relu = tf.nn.relu(conv2)
    pool2 = tf.nn.max_pool(conv2_relu, ksize=[1, 2, 2, 1],
strides=[1, 2, 2, 1], padding='VALID')


    # Flatten the output from convolutional layers
    flattened = tf.keras.layers.Flatten()(pool2)
```

(continued)

Listing 6.10 (continued)

```
    # Fully Connected Layer
    fc = tf.matmul(flattened, fc_W) + fc_b
    fc_relu = tf.nn.relu(fc)

    # Output Layer
    logits = tf.matmul(fc_relu, out_W) + out_b
    output = tf.nn.softmax(logits)


    return output
```

6.6.4 *Training*

The following is an implementation of training a CNN model, leveraging the forward propagation code as we have implemented before. Since we are dealing with a multi-class classification problem, we will be utilizing categorical cross-entropy loss (Bishop and Nasrabadi 2006) here:

Listing 6.11 Training a CNN Model

```
def loss_function(y_hat, y):
    loss = -tf.reduce_sum(y * tf.math.log(y_hat), axis=-1)
    loss = tf.reduce_mean(loss)

    return loss

def train(X_train, y_train, parameters, num_epochs=10,
learning_rate=0.001):

    # Define the optimizer
    optimizer = tf.keras.optimizers.legacy.Adam(learning_rate)

    # Training loop
    for epoch in range(num_epochs):
        with tf.GradientTape() as tape:
            tape.watch(parameters)
            logits = forward_propagation(X_train, parameters)
            loss = loss_function(logits, y_train)
```

(continued)

Listing 6.11 (continued)

```
        gradients = tape.gradient(loss, parameters.values())
        optimizer.apply_gradients(zip(gradients, parameters.
values()))

        if epoch % 100 == 0:
            print('Epoch: %d, Loss: %f' % (epoch, loss))

    return parameters
```

In this training function, we are reusing our previously implemented `forward_propagation` code. We then utilize TensorFlow's `tape.gradient` to compute the gradients of the loss with respect to the parameters. The optimizer applies these gradients to update the parameters, iterating over the specified number of epochs.

We will now train the citrus disease classification model for 500 epochs as follows:

```
input_shape = (28, 28, 3)
num_classes = 4
parameters = initialize_parameters(input_shape, num_classes)
parameters = train(X_train, y_train, parameters, num_epochs=500,
learning_rate=0.001)

Epoch: 0, Loss: 1.677086
Epoch: 100, Loss: 0.290670
Epoch: 200, Loss: 0.095202
Epoch: 300, Loss: 0.033464
Epoch: 400, Loss: 0.014253
```

6.6.5 *Prediction and Validation*

As we have previously discussed, prediction essentially involves a single pass of forward propagation using the trained parameters. This process generates a prediction probability for each class. The class with the highest probability is selected as the final prediction. The code snippet provided below demonstrates the use of our predictive model on the test set:

```
import numpy as np
y_pred_proba = forward_propagation(X_test, parameters)
y_pred_proba = y_pred_proba.numpy()
y_pred = np.argmax(y_pred_proba, 1)
```

Let us now work on the validation and generate a classification report (Pedregosa et al. 2011):

```
from sklearn.metrics import classification_report
print(classification_report(y_test.numpy().argmax(axis=1), y_pred))
```

```
              precision    recall  f1-score   support

           0       0.76      0.76      0.76        29
           1       0.88      0.93      0.90        30
           2       0.83      0.83      0.83        46
           3       0.92      0.79      0.85        14

    accuracy                           0.83       119
   macro avg       0.84      0.83      0.83       119
weighted avg       0.83      0.83      0.83       119
```

The classification report reveals that the F1-score of our image classification model stands at 0.83. This is a remarkable achievement. We have successfully built an image classification model from scratch, utilizing convolutional neural networks.

6.7 Summary

- Convolutional neural networks (CNNs) are designed primarily for computer vision applications such as image recognition and classification tasks.
- CNNs are structured with convolutional layers that apply filters to input data, pooling layers for down-sampling, fully connected layers for feature aggregation, and activation functions for introducing nonlinearity.
- Filters (kernels) in convolutional layers detect patterns such as edges, textures, and shapes by sliding over the input to create feature maps.
- The stride defines the step size for the filter movement, and padding adds zeros around the input to maintain spatial dimensions.
- Pooling layers reduce the dimensionality of feature maps by down-sampling, commonly using max pooling or average pooling methods, which help in reducing computational complexity and preventing overfitting.
- Fully connected layers, present at the end of CNN architectures, leverage the features extracted by convolutional and pooling layers to make predictions.

References

Bishop, C. M., & Nasrabadi, N. M. (2006). *Pattern recognition and machine learning* (Vol. 4). Springer.

Chollet, F. (2021). *Deep learning with Python*. Simon and Schuster.

Dumoulin, V., & Visin, F. (2016). A guide to convolution arithmetic for deep learning. *arXiv Preprint arXiv:1603.07285*.

Goodfellow, I. (2016). *Deep learning*. MIT press.

Gu, J., Wang, Z., Kuen, J., Ma, L., Shahroudy, A., Shuai, B., Liu, T., Wang, X., Wang, G., Cai, J., & others. (2018). Recent advances in convolutional neural networks. *Pattern Recognition*, *77*, 354–377.

He, K., Zhang, X., Ren, S., & Sun, J. (2015). Delving deep into rectifiers: Surpassing human-level performance on imagenet classification. *Proceedings of the IEEE International Conference on Computer Vision*, 1026–1034.

Krizhevsky, A., Sutskever, I., & Hinton, G. E. (2012). Imagenet classification with deep convolutional neural networks. *Advances in Neural Information Processing Systems*, *25*.

LeCun, Y., Bottou, L., Bengio, Y., & Haffner, P. (2002). Gradient-based learning applied to document recognition. *Proceedings of the IEEE*, *86*(11), 2278–2324.

Marr, D., & Hildreth, E. (1980). Theory of edge detection. *Proceedings of the Royal Society of London. Series B. Biological Sciences*, *207*(1167), 187–217.

Nair, V., & Hinton, G. E. (2010). Rectified linear units improve restricted boltzmann machines. *Proceedings of the 27th International Conference on Machine Learning (ICML-10)*, 807–814.

Pedregosa, F., Varoquaux, G., Gramfort, A., Michel, V., Thirion, B., Grisel, O., Blondel, M., Prettenhofer, P., Weiss, R., Dubourg, V., & others. (2011). Scikit-learn: Machine learning in Python. *The Journal of Machine Learning Research*, *12*, 2825–2830.

Rauf, H. T., Saleem, B. A., Lali, M. I. U., Khan, M. A., Sharif, M., & Bukhari, S. A. C. (2019). A citrus fruits and leaves dataset for detection and classification of citrus diseases through machine learning. *Data in Brief*, *26*, 104340.

Scherer, D., Müller, A., & Behnke, S. (2010). Evaluation of pooling operations in convolutional architectures for object recognition. *International Conference on Artificial Neural Networks*, 92–101.

Simonyan, K., & Zisserman, A. (2014). Very deep convolutional networks for large-scale image recognition. *arXiv Preprint arXiv:1409.1556*.

Yokoya, N., Grohnfeldt, C., & Chanussot, J. (2017). Hyperspectral and multispectral data fusion: A comparative review of the recent literature. *IEEE Geoscience and Remote Sensing Magazine*, *5*(2), 29–56.

Zeiler, M. D., & Fergus, R. (2014). Visualizing and understanding convolutional networks. *Computer Vision–ECCV 2014: 13th European Conference, Zurich, Switzerland, September 6-12, 2014, Proceedings, Part I 13*, 818–833.

Chapter 7
Creating Recurrent Neural Networks

This chapter covers:

- Introducing recurrent neural network (RNN) representation in sequence modeling tasks.
- Understanding tokenization technique for text processing.
- Exploring forward propagation with RNN cells.
- Overcoming challenges in gradient descent in sequence models.
- Implementing text generation methods with RNN.
- Demonstrating a case study: creative text generation.

In this chapter, we begin with an explanation of why recurrent neural networks (RNNs) are needed and how they handle sequential data. We explore the basics of tokenization—covering character-level approach. We then dive into the mechanics of RNN forward propagation, exploring how these networks process sequential information iteratively. Throughout, we address practical aspects such as loss computation, gradient propagation challenges such as vanishing and exploding gradients, and techniques for mitigating them. Additionally, we look into text generation using RNNs and compare different sampling and decoding methods. This chapter aims to provide a clear foundation and practical insights into leveraging RNNs effectively, supported by examples and code snippets.

7.1 RNN Representation

Tasks related to sequence modeling deal with data that is sequential in nature. The key idea behind sequence modeling tasks is that the output for a given input element depends not only on that element but also on a series of preceding elements. Some of the common sequence modeling tasks are:

T. Islam, *Hands-on Deep Learning*, https://doi.org/10.1007/978-3-032-00488-8_7

- Machine translation
- Speech recognition
- Named entity recognition (NER)
- Question and answering
- Text classification
- Sentiment classification
- Creative text and music generation
- Time series forecasting

However, traditional neural networks such as fully connected or convolutional neural networks fall short when it comes to sequence modeling tasks. This is because they fail to account for the sequential nature of the data, which is a critical aspect of these tasks. They are unable to grasp the sequence-based relationships, which makes them unsuitable for tasks where the sequence of inputs holds substantial importance.

Recurrent neural networks (RNNs) are a class of neural networks that excel in processing sequential data. Unlike traditional neural networks, RNNs possess a form of memory that allows them to maintain information over time. This is achieved through the use of hidden states, which are updated at each time step of the sequence and carry information from previous steps. The hidden state at a given time step is a function of the current input and the previous hidden state. This recursive process enables RNNs to efficiently understand time-related dependencies in the data, making them especially appropriate for tasks such as natural language processing, time series analysis, and more.

To illustrate the representation of recurrent neural networks (RNNs) in sequence modeling, let us consider the task of named entity recognition (Collobert et al. 2011). This task involves identifying named entities in a sentence, such as a person, location, or time. For instance, consider the sentence "Taylor Swift performed in Washington in 2023." The task is to classify first and second tokens as Person, fifth token as location, and last token as time entity. This is illustrated in Fig. 7.1. This task can be effectively modeled using an RNN. At each time step, the RNN updates its hidden state based on the current input token and the previous hidden state. This mechanism allows it to retain a form of memory about the sequence of tokens encountered so far. The final output of the RNN is the classification of each token into an entity type. This is an example of a many-to-many RNN architecture.

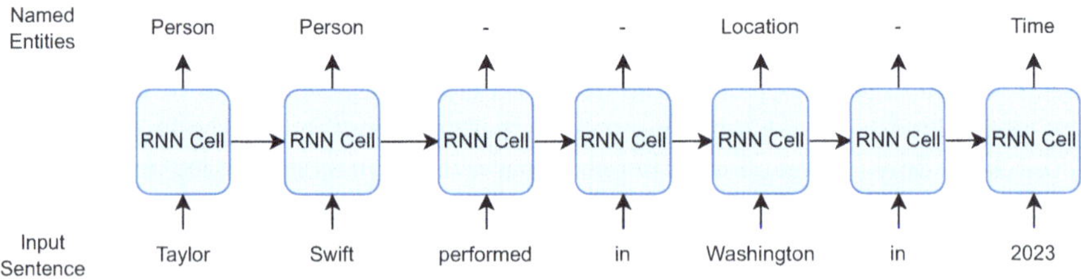

Fig. 7.1 An example illustrating named entity recognition task from text

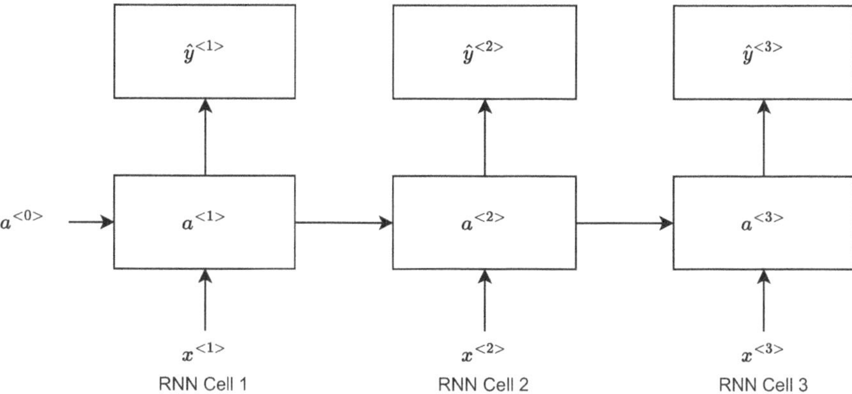

Fig. 7.2 RNN representation in time steps

While "Washington" could be interpreted as a "Person," in the given context, it is a "Location" entity. This is why we will need sequence models such as RNNs. These models can understand the context within a sequence, which is crucial for accurate entity classification. The ability of RNNs to maintain information about previous inputs in their hidden state allows them to interpret "Washington" as a location, based on its position and role in the sentence, rather than just its standalone meaning. This context-aware processing is what makes RNNs particularly effective for tasks such as named entity recognition.

To dive deeper into the workings of a recurrent neural network (RNN), let us consider a many-to-many architecture, as depicted in Fig. 7.2. At the first time step, the RNN reads the input $x^{<1>}$, with information flowing from the previous time step, which in this case is the initial activation $a^{<0>}$, initialized to zeros. A new activation $a^{<1>}$ is computed at this time step using the inputs $x^{<1>}$ and $a^{<0>}$. This activation is then passed through a sigmoid or softmax activation function to compute the output $\widehat{y}^{\langle 1 \rangle}$ for the first time step. This process is repeated for each subsequent time step, calculating $a^{\langle 2 \rangle}, \widehat{y}^{\langle 2 \rangle}, a^{\langle 3 \rangle}, \widehat{y}^{\langle 3 \rangle}$, and so on. The key point to note here is that the activation at each time step is influenced not only by the input at that time step but also by the activations from previous time steps. This allows the RNN to maintain a form of temporal memory, which is crucial for effectively handling sequence modeling tasks.

In the context of sequence modeling, T_x and T_y are used to denote the lengths of the input and output sequences, respectively. Here is a bit more detail:

- T_x: This represents the length of the input sequence. In other words, it is the number of time steps that the network sees input data.
- T_y: This is the length of the output sequence or the number of time steps that the network produces output data.

These parameters are crucial when dealing with sequences of varying lengths, and they can differ depending on the task. Figure 7.3 illustrates a variety of RNN

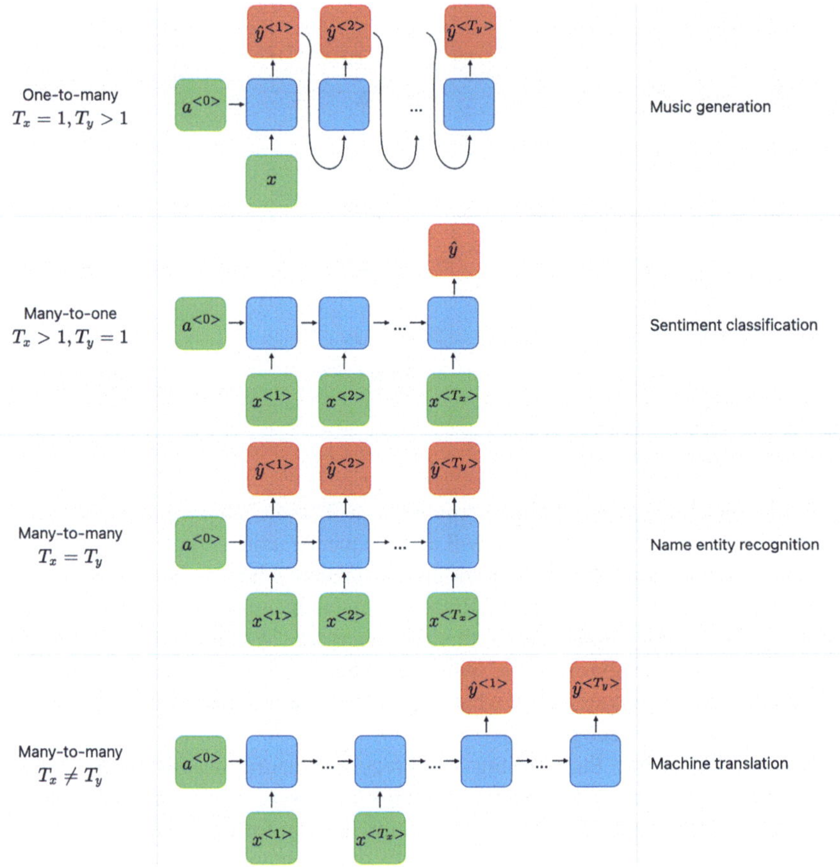

Fig. 7.3 Common RNN architectures and use cases

architectures, each tailored for specific tasks. For instance, in our named entity recognition example, $T_x = T_y$. On the other hand, in sequence-to-sequence tasks such as machine translation (Cho et al. 2014a; Sutskever et al. 2014), T_x and T_y can be different as the length of the source sentence can be different from the length of the target sentence. This is also known as many-to-many RNN architecture, but follows encoder–decoder architecture (Bahdanau et al. 2014). In image captioning tasks (Karpathy and Fei-Fei 2015), where a single input leads to a sequence of outputs, the T_x would be 1 and T_y would be greater than 1. This is known as one-to-many RNN architecture. In sentiment analysis tasks, a sequence of input tokens leads to a single output, thus T_x would be greater than 1 and T_y would be 1. This is known as many-to-one RNN architecture. Creative text generation or music generation follows one-to-many RNN architecture. Here, the input is typically a single seed or trigger, such as a starting token, thus $T_x = 1$. This input is processed by the RNN, which then generates a sequence of outputs. In the case of text generation, the

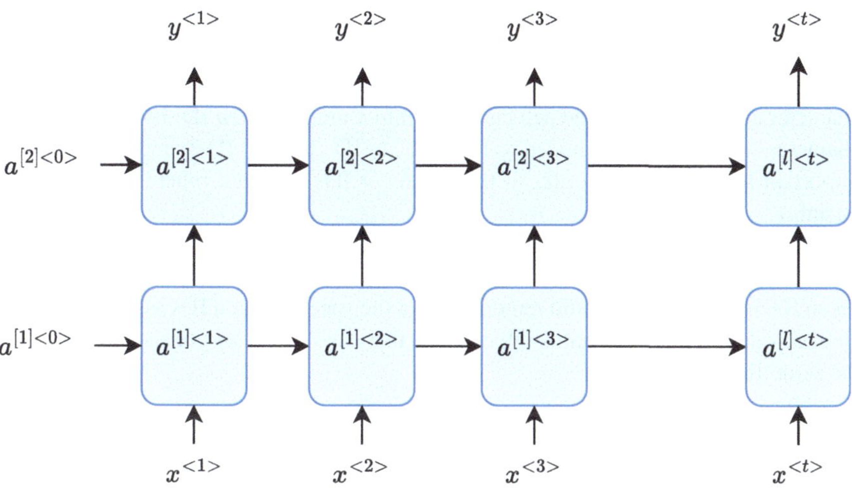

Fig. 7.4 Illustration of deep RNNs by stacking RNN layers

output could be a sentence, paragraph, or even an entire story. The one-to-many architecture of the RNN allows it to take this single input and create a diverse and creative array of outputs, making it an ideal choice for tasks involving creative generation. The RNN's ability to remember past inputs in its hidden state also allows it to generate outputs that are contextually relevant and coherent. We will go through a step-by-step example in the latter part of the chapter.

Until now, our discussion has been focused on a single layer of recurrent neural networks (RNNs). However, it is important to note that these RNN layers can be stacked atop one another, thereby creating a more complex and powerful architecture known as Deep RNNs (Graves et al. 2013; Jozefowicz et al. 2015; Pang et al. 2019). This concept is visually represented in Fig. 7.4.

There is another RNN representation worth mentioning. This is called Bidirectional RNN (Bi-RNN) (Schuster and Paliwal 1997). Unlike standard RNNs, which process data sequentially from the beginning to the end, Bi-RNNs are designed to interpret the data in both directions. For the sake of simplicity, this chapter will be dedicated to the construction of a single layer of a Unidirectional RRN.

7.2 Tokenization

Tokenization is a crucial step in RNN, as neural networks are not capable of understanding or performing operations on raw strings or text. Tokenization involves breaking down natural language into distinct parts, which are then converted into token IDs. This process allows us to transform human-readable text into a format that can be understood and processed by a neural network. There are several types of tokenization, including character-level, word-level, and sub-word tokenization (Sennrich et al. 2015).

Character-level tokenization, for example, involves breaking down a text into character and then encoding each character into a unique token ID. For instance, the character "a" might be tokenized to ID 1, "b" to ID 2, "c" to ID 3, and so forth. It is important to note that the order of the tokens or IDs does not inherently carry any meaning.

Here is an example of how one might code up a character-level tokenizer. This code demonstrates the basic concept of mapping each unique character to a distinct token ID. However, we should remember that the specific token IDs assigned to each character do not matter; what is important is that each character consistently maps to the same ID:

Listing 7.1 Tokenizer

```
class Tokenizer:
    def __init__(self):
        self.char_index = {}
        self.index_char = {}

    def fit(self, text):
        # Unique chars
        unique_chars = sorted(set(text))

        # Create a dictionary mapping each character to a unique index
        self.char_index = {char: index for index, char in enumerate
(unique_chars)}

        # Create a dictionary mapping each unique index to a character
        self.index_char = {index: char for index, char in enumerate
(unique_chars)}

    def encode(self, text):
        # Encode the text
        encoded_ids = [self.char_index.get(char, 0) for char in text]
        return encoded_ids

    def decode(self, encoded_ids):
        # Decode the text
        decoded_text = ''.join(self.index_char.get(index, '') for
index in encoded_ids)
        return decoded_text
```

7.3 RNN Forward Propagation

7.3.1 Single RNN Cell Step

Let us now walk through a single RNN cell step for many-to-many scenario. The operations for a single RNN cell step is described in Fig. 7.5. The RNN cell takes two inputs: the current input, denoted as $x^{\langle t\rangle}$, and the previous hidden state, $a^{\langle t-1\rangle}$, which encapsulates information from the past. The output of this cell is $a^{\langle t\rangle}$, which is passed on to the next RNN cell and is also utilized to predict $y^{\langle t\rangle}$.

Here are the main two operations in a single RNN cell step:

1. First, we compute the hidden state using the tanh activation function. This can be represented mathematically as:

$$a^{\langle t\rangle} = \tanh\left(W_{aa}a^{\langle t-1\rangle} + W_{ax}x^{\langle t\rangle} + b_a\right)$$

2. Next, we use the newly computed hidden state, $a^{\langle t\rangle}$, to calculate the prediction

$$\hat{y}^{\langle t\rangle} = softmax\left(W_{ya}a^{\langle t\rangle} + b_y\right)$$

This means that learning process involves the optimization of several parameters. These include the weight matrices W_{aa}, W_{ax}, W_{ya}, as well as the bias vectors b_a and b_y. The weight matrix W_{aa} is responsible for transforming the previous hidden state $a^{\langle t-1\rangle}$, while W_{ax} is used for the current input $x^{\langle t\rangle}$. Together, they contribute to the

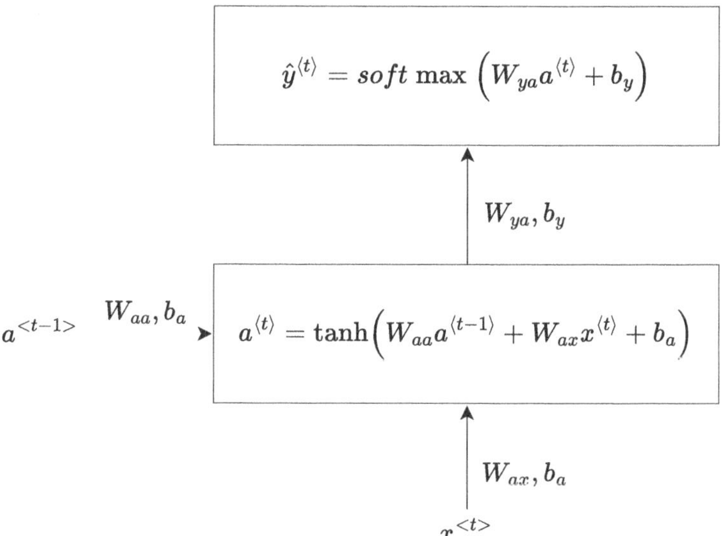

Fig. 7.5 A detailed view of a single RNN cell step and its associated operations

computation of the current hidden state, $a^{\langle t \rangle}$. The bias vector b_a is added to this computation, providing a degree of flexibility to the model. On the other hand, W_{ya} is the weight matrix that transforms the current hidden state, $a^{\langle t \rangle}$, into the output prediction, $\widehat{y}^{\langle t \rangle}$. The bias vector b_y is added to this computation, again adding flexibility. Note that lowercase notation is typically used to represent RNNs, even though these notations can also represent matrices. We will adhere to this convention in our discussion.

Following code initializes the parameters using He initialization for RNN:

Listing 7.2 Initialization of RNN Parameters

```
def initialize_parameters(n_x, n_a, n_y, initialization="he"):
    tf.random.set_seed(210)
    if initialization == "he":
        Wax = tf.Variable(tf.random.normal([n_a, n_x]) * tf.sqrt
(2. / n_x), trainable=True)
        Waa = tf.Variable(tf.random.normal([n_a, n_a]) * tf.sqrt
(2. / n_a), trainable=True)
        Wya = tf.Variable(tf.random.normal([n_y, n_a]) * tf.sqrt
(2. / n_a), trainable=True)
    else:
        print("Not Implemented")

    ba = tf.Variable(tf.zeros([n_a, 1]), trainable=True)
    by = tf.Variable(tf.zeros([n_y, 1]), trainable=True)

    parameters = {"Wax": Wax,
                  "Waa": Waa,
                  "Wya": Wya,
                  "ba": ba,
                  "by": by}

    return parameters
```

In the above code:

- n_x: This represents the number of features or the size of the vocabulary in the input $x^{\langle t \rangle}$.
- n_y: This is the number of features or the size of the vocabulary in the output $y^{\langle t \rangle}$.
- n_a: This denotes the number of units in the hidden state $a^{\langle t \rangle}$ of the RNN.

The single RNN cell step will then look like this:

Listing 7.3 Single RNN Cell Step

```
def rnn_single_cell_step_forward(x_t, a_t_minus_1, parameters):
    Wax = parameters["Wax"]
    Waa = parameters["Waa"]
    Wya = parameters["Wya"]
    ba = parameters["ba"]
    by = parameters["by"]

    a_t = tf.tanh(tf.matmul(Waa, a_t_minus_1) + tf.matmul(Wax,
x_t) + ba)
    y_t_hat = tf.nn.softmax(tf.matmul(Wya, a_t) + by, axis=0)

    return a_t, y_t_hat
```

7.3.2 Combining Cell Steps

The process of forward propagation in recurrent neural networks (RNNs) can be thought of as a series of repeated single RNN cell steps. This iterative process is what allows RNNs to handle sequential data effectively. Let us revisit our example sentence to illustrate this: "*Taylor Swift performed in Washington in 2023*." In the context of named entity recognition (NER), our objective is to predict the entity type for each word in the sentence. Each word represents a time step in the sequence, and the entity prediction for each word is the outcome of that time step. At the first time step, the RNN processes the word "Taylor." The model, based on its learned parameters and the activation function applied to the input, aims to predict that "Taylor" is part of a "Person" entity. In the subsequent time step, the model processes the word "Swift." Ideally, the model should predict that "Swift" is also part of the "Person" entity, thereby recognizing "Taylor Swift" as a complete name. This process continues, with each time step involving the prediction of the entity for the current word in the sequence, until the entire sentence has been processed.

One of the key strengths of RNNs is their ability to utilize context from previous time steps to inform predictions at the current time step. For instance, when the model processes the word "Washington," it has already encountered the phrase "Taylor Swift performed in." With this context, the model should be able to accurately predict that "Washington" is a "Location" entity, rather than a "Person" entity. In summary, forward propagation in RNNs involves a series of repeated steps performed by individual RNN cells, each processing a word in the sequence and predicting its entity type, while leveraging the context from previous time steps. One may refer to Fig. 7.2 once again for further intuition. This mechanism allows RNNs to effectively handle sequential data and perform tasks such as named entity recognition.

Let us now go over the dimensions of the inputs, outputs, and parameters involved in the forward propagation of a recurrent neural network (RNN). Table 7.1 provides a clear overview of the dimensions of various elements involved

Table 7.1 Overview of the dimensions of various elements involved in an RNN's forward propagation

Parameter	Description	Dimensions
x	Input matrix across all time steps	(n_x, m, T_x)
W_{ax}	Weight matrix associated with the input	(n_a, n_x)
W_{aa}	Weight matrix associated with the hidden state	(n_a, n_a)
W_{ya}	Weight matrix associated with the output	(n_y, n_a)
b_a	Bias vector associated with the hidden state	(n_a, 1)
b_y	Bias vector associated with the output	(n_y, 1)
a	Hidden state across all time steps	(n_a, m, T_x + 1)
y	Predictions across all time steps	(n_y, m, T_y)

in an RNN's forward propagation. It is important to understand these dimensions for implementing and debugging RNNs effectively.

Let us now implement the forward propagation of the recurrent neural network (RNN) by integrating each individual RNN cell step:

Listing 7.4 RNN Forward Propagation

```
def rnn_forward_propagation(x, parameters):
    n_x, m, T_x = x.shape
    n_y, n_a = parameters["Wya"].shape

    a = tf.TensorArray(dtype=tf.float32, size=T_x + 1)
    y_hat = tf.TensorArray(dtype=tf.float32, size=T_x)

    a0 = tf.zeros([n_a, m])
    a = a.write(0, a0)

    for t in tf.range(1, T_x + 1):
        a_t_minues_1 = a.read(t - 1)
        a_t, y_t_hat = rnn_single_cell_step_forward(x[:, :, t - 1],
a_t_minues_1, parameters)
        a = a.write(t, a_t)
        y_hat = y_hat.write(t - 1, y_t_hat)

    # Ensure that the final tensors have the correct shapes
    a = a.stack()
    y_hat = y_hat.stack()

    a = tf.transpose(a, [1, 2, 0])
    y_hat = tf.transpose(y_hat, [1, 2, 0])

    return a, y_hat
```

Here, we start by initializing two arrays, one to store the activations over time and the other to hold the predicted outputs. We then set the initial activation to zero. Next, we enter a loop that iterates over each time step. In each iteration, we read the activation from the previous time step, and use it along with the current input to compute the current activation and predicted output. These values are then stored in their respective arrays. After iterating through all the time steps, we stack the activations and predicted outputs to ensure they have the correct shapes. Finally, we transpose these stacked tensors to match the original input shape, and return them as the result of the forward propagation. This process allows us to capture and process temporal dependencies in our input data through the RNN.

Let us consider another example related to next token prediction. Suppose we are trying to predict the letters in the word "umbrella", as part of an autocomplete system. Each letter prediction is considered a time step in the sequence. In the first time step, we start with the letter "u" and aim to predict the next letter "m." The RNN cell takes "u" as input and outputs a prediction for the next character. This prediction is based on the learned parameters and the activation function applied to the input. In the second time step, the RNN cell takes the actual next character "m" and the hidden state from the previous time step as inputs. It then predicts the following character "b." This process continues, with each time step predicting the next character given the current one, until we have moved through the entire word "umbrella."

So, basically what we are doing here is when predicting "a," we have all the information from previous steps: "umbrell." So, we are basically computing *P (current token | previous tokens)*. This is the essence of RNN forward propagation. It is a powerful mechanism that allows the network to maintain a form of "memory" about previous inputs in the sequence, which is crucial for tasks such as language modeling (Mikolov et al. 2010), text generation, and many other applications involving sequential data. It is fascinating to see how these simple steps, when repeated, can lead to such complex and useful behaviors in the network.

7.4 RNN Backpropagation

7.4.1 Gradients

We know that the computation of gradients depends on the loss function. The first step in RNN backpropagation is to define the loss function prior to initiating the backpropagation process. Let us consider a scenario where we are interested in next token prediction task, as previously discussed. In this setup, each time step involves a multi-class classification task. Thus, the most suitable loss function for this scenario is the categorical cross-entropy loss (Bishop and Nasrabadi 2006; Goodfellow 2016). This loss function is particularly effective in situations where the model needs to make decisions among multiple categories.

The loss for each time step t can be calculated using the negative log-likelihood of the true label y given the prediction $\widehat{y}$. This is often used in classification problems and is defined as follows.

For a single data point and for time step t:

$$L_t = -\sum_c y_{t,c} \log\left(y_{\hat{t},c}\right)$$

In the context of a sequence, the total loss would be the average of the losses at each time step. If we have T time steps, it can be written as:

$$L = \frac{1}{T}\sum_{t=1}^{T} L(y_t, \widehat{y}_t)$$

The cost function is then averaged over the examples:

$$J = \frac{1}{m}\sum_{i=1}^{m} L$$

where:

- $L(y_t, \widehat{y}_t)$ is the loss at time step t
- $y^{\langle t \rangle}$ is the true label at time step t
- $\widehat{y}^{\langle t \rangle}$ is the predicted label at time step t
- J is the cost function (average loss over all examples)
- m is the number of examples in the batch
- T is the total number of time steps.

Let us now code the loss function:

Listing 7.5 Loss over Timesteps

```
def loss_function(y_hat, y):

    # Compute loss at each timestep for each example (for n_classes/
one hot)
    loss = -tf.reduce_sum(y * tf.math.log(y_hat), axis=0)

    # Average over timesteps for each example
    loss = tf.reduce_mean(loss, axis=-1)

    # Average over examples
    loss = tf.reduce_mean(loss)

    return loss
```

For the backpropagation, we can use TensorFlow's GradientTape automatic differentiation. Recall, we have already used this in the previous chapter. TensorFlow's GradientTape is a context manager that allows us to compute gradients. It records operations for automatic differentiation. When performing backpropagation, we can retrieve the gradients of any tensor with respect to any other tensor. A Python implementation will look like this:

```
# Target y
y = tf.random.uniform(shape=[n_y, m, T_y], minval=0, maxval=28)

# Use tf.GradientTape for automatic differentiation
with tf.GradientTape() as tape:
tape.watch(parameters) # Watch parameters

# Forward propagation
a, y_hat = rnn_forward_propagation(x, parameters)

# Loss
loss = loss_function(y, y_hat)

# Get the gradients
gradients = tape.gradient(loss, parameters.values())
```

7.4.2 Vanishing and Exploding Gradients

We have already covered vanishing and exploding gradients concepts before. We have also discussed that He initialization is a technique that can help mitigate the vanishing and exploding gradient problems. However, while it can help to some extent, it does not fully solve the problem, especially in the context of RNNs and long sequences. The vanishing gradients and exploding gradients problems become particularly pronounced when dealing with longer sequences (Bengio et al. 1994; Pascanu et al. 2013).

Vanishing Gradients As the sequence length increases, gradients passed back in time may decrease exponentially, eventually becoming insignificantly small (i.e., they "vanish"). This is problematic because it means that the RNN cannot learn to connect information from long ago to the present task. For example, if we are trying to predict the next word in a sentence, an RNN with vanishing gradients might not be able to remember enough about the beginning of the sentence to make accurate predictions at the end. This makes it hard for an RNN to learn long-term dependencies (Bengio et al. 1994).

Exploding Gradients The opposite problem can also occur, where gradients grow exponentially during backpropagation. This leads to large updates to neural network parameters and can result in an unstable network. At an extreme, the values of weights can become so large and result in NaN values.

The core reason behind these problems is the use of the chain rule for backpropagation (LeCun et al. 2002; Werbos 1994) in RNNs. When the sequence is long, we multiply many gradients together during backpropagation. If these gradients are small (less than 1), the result can vanish to zero. Conversely, if they are large (greater than 1), the result can explode.

There are several methods to mitigate these problems:

- *Gradient clipping*: This is a technique that prevents exploding gradients by artificially limiting (or "clipping") them to a maximum value (Chen et al. 2020; Zhang et al. 2019). This can help prevent large updates to the model's parameters.
- *Long short-term memory (LSTM) and gated recurrent unit (GRU)*: These are unique variants of recurrent neural networks (RNNs) (Cho et al. 2014b; Hochreiter and Schmidhuber 1997). They incorporate gating mechanisms within their structure to regulate the information flow. This mechanism aids in alleviating the vanishing gradient issue by establishing paths where the gradient can flow without becoming too small. The upcoming chapter will cover into the details of LSTM.

Gradient clipping is straightforward to implement. In the gradient clipping method, we define a threshold value. If the magnitude of a gradient vector surpasses this threshold value, we clip the gradient to the threshold value. Here is the code:

Listing 7.6 Gradient Clipping

```
def gradient_clipping(gradients, max_value):

    clipped_gradients = [tf.clip_by_value(grad, -max_value,
max_value) for grad in gradients]

    return clipped_gradients
```

7.5 Text Generation

In certain sequence modeling tasks such as sentiment classification and named entity recognition, there is no need to generate text. Instead, the goal is to classify entities or sentiments. However, when it comes to tasks that involve text generation, such as creative text generation or machine translation, the primary task is to generate text using a trained recurrent neural network (RNN) model.

For text generation, we often use a technique called "sampling" (Holtzman et al. 2019) instead of simply choosing the word with the maximum probability predicted by the recurrent neural network (RNN) at each step. This approach introduces an element of randomness into the text generation process, which can lead to more diverse and creative outputs. The temperature hyperparameter dictates the degree of randomness we desire in the model's output. By using sampling and adjusting the temperature, we can control the trade-off between randomness (creativity) and predictability (accuracy). This is particularly useful in creative text generation tasks where we want the output to be not just grammatically correct but also interesting and diverse.

RNN actually predicts a probability distribution over the vocabulary for the next token (e.g., character or word). This means that for each possible word, the RNN assigns a probability that this word is the next word in the sequence. Instead of choosing the word with the highest probability—which is called greedy decoding (Gu et al. 2017), we can sample from this probability distribution. This means that a word with a lower probability also has a chance to be selected, albeit with low probability. The word to be selected is not always the one with the highest probability, but is chosen based on the probability distribution. We can control the randomness of the sampling process by adjusting a parameter called the temperature. A higher temperature results in more randomness, while a lower temperature makes the distribution sharper, meaning the word with the highest probability is more likely to be chosen.

Here are some basic guidelines for setting temperature values during sampling:

- *Temperature* $= 1$: When the temperature is 1, the model uses the original probabilities predicted by the model. This means that each word is chosen with a probability proportional to its predicted probability, leading to a balance between diversity and coherence.
- *Temperature* >1: When the temperature is greater than 1, it makes the probability distribution "flatter." This means that less probable words are given a higher chance of being selected, which increases diversity but can also lead to more randomness and less coherence.
- *Temperature* <1: When the temperature is less than 1, it makes the probability distribution "sharper." This means that the most probable words are given an even higher chance of being selected, which decreases diversity and makes the output more deterministic and coherent.
- *Temperature* $= 0$: When the temperature is 0, it is equivalent to greedy decoding. This decoding strategy always selects the word with the highest probability.

In Fig. 7.6, we illustrate the impact of varying the temperature parameter on the sampling process. This is achieved by plotting the probabilities, which are assumed to follow a Gaussian distribution, under different sampling temperatures. For the purpose of this visualization, we make two assumptions: first, that our vocabulary comprises 1000 tokens, and second, that the peak probabilities are primarily concentrated around the token ID of 500. So, temperature sampling should select the token near this token ID based on the visualization. When the temperature parameter

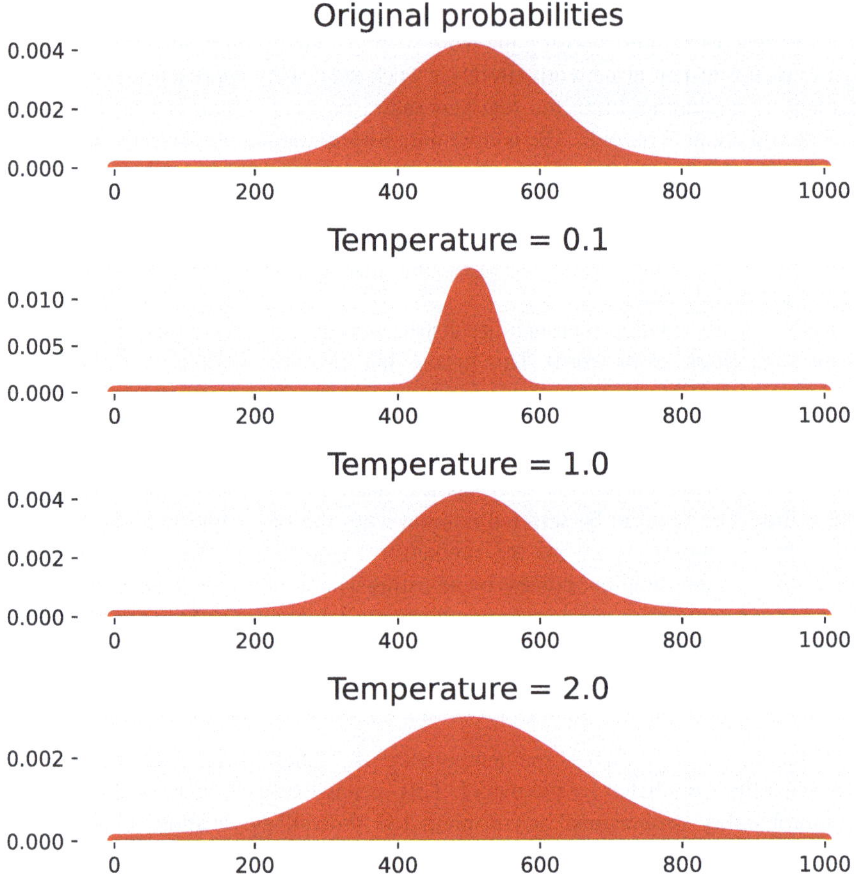

Fig. 7.6 The impact of varying the temperature parameter on the sampling

is set to 0.1, the selection region becomes more concentrated, thereby diminishing the probability of selecting a token with an ID around 200. On the contrary, when the temperature parameter is increased to 2.0, the selection region broadens, thereby enhancing the likelihood of more frequently selecting tokens with an ID in the vicinity of 200.

Building on the concepts discussed above, we now codify the text generation for a creative text generation use case:

Listing 7.7 Generating Text

```
def generate_text(parameters, seed, temperature=1.0):
    n_a, n_x = parameters["Wax"].shape
    T_x = 20

    # Tokenizer ids
    ids = tokenizer.index_char.keys()
    ids = np.array(list(ids), dtype=int)

    a_t_minus_1 = tf.zeros((n_a, 1))

    idx = tokenizer.encode('<')[0]
    x_t = tf.zeros((n_x, 1))
    x_t = tf.tensor_scatter_nd_update(x_t, [[idx]], [[1]])

    indices = []
    text = ''

    for i in range(T_x):
        a_t_minus_1, y_t_hat = rnn_single_cell_step_forward(x_t,
a_t_minus_1, parameters)

        # Convert y_pred to a numpy array
        p = np.array(np.squeeze(y_t_hat))

        # Apply temperature adjustment
        p = np.exp(np.log(p) / temperature)

        # Normalize probabilities to ensure they sum up to 1
        p /= p.sum()

        # Sample an index based on the probability distribution
        idx = np.random.choice(ids, p=p)

        # Update x
        x_t = tf.one_hot([idx], depth=n_x, axis=0, dtype=tf.float32)

        # Decode char
        decoded_char = tokenizer.decode([idx])
        if decoded_char == '>':
            break
        text = text + decoded_char

return text
```

We often initiate the text generation process with a specific start token (in this case, “<”). This token serves as the initial input to our text generation model. The model then begins to decode or generate text, token by token, based on the learned patterns from the training data. The model continues to generate text until it encounters a specific end token. This end token signifies the completion of a text sequence. The “temperature” parameter controls the randomness in the model’s output. We will see an example of creative text generation case study next.

7.6 Building an End-to-End RNN Model

With our foundational knowledge now complete—from text tokenization to the intricacies of RNN forward and backpropagation—we possess all the essential building blocks to construct a full end-to-end RNN model. It is time to put this theoretical understanding into practical application, exploring a compelling real-world use case: creative text generation.

7.6.1 Creative Text Generation Problem

In this case study, we will be exploring a dataset of family names from the Angiosperms, also known as flowering plants, which belong to the kingdom “Plantae.” There are approximately 412 known families of Angiosperms. Now, imagine a scenario where a new family is discovered. We are then tasked to devise a name for this new family. The task here is to train a recurrent neural network (RNN) model to learn the patterns inherent in these family names and generate creative names for potential new Angiosperms families. It is noteworthy that the family names of Angiosperms typically end with the suffix “-aceae,” with a few exceptions for certain historic names. We would like to see if the RNN can learn these patterns and generate names that bear a resemblance to the existing ones.

Let us create loading utility first. We will also add start and end token in each family name:

Listing 7.8 Loading Angiosperms Data

```
def load_angiosperms_data():

    with open("../data/Angiosperms.txt") as f:
    data = f.readlines()
    data = [x.lower().strip() for x in data]
    data = ['<' + word + '>' for word in data]

    return data
```

Here are some examples of family names included in our dataset, when we load the data:

```
data = load_angiosperms_data()
data[0:5]
```

```
['<acanthaceae>',
 '<achariaceae>',
 '<achatocarpaceae>',
 '<acoraceae>',
 '<actinidiaceae>']
```

7.6.2 *Sequence Preparation*

Next step is sequence preparation. This includes tokenization, that is converting each token into token ID, and adding padding. This can be achieved with the following code:

```
combined_text = ' '.join(data)
tokenizer = Tokenizer()
tokenizer.fit(combined_text)
```

The subsequent step in our process involves preparing the sequences. This process encompasses two key tasks: tokenization and padding. Tokenization is the process of converting each token into a unique identifier, or token ID. This allows us to represent our data in a format that our model can understand and learn from. Padding, on the other hand, is used to ensure that all sequences are of the same length.

In the provided code below, we define two functions:

Listing 7.9 Encoding, Padding, and Preparing Sequences

```
from tensorflow.keras.utils import to_categorical
from tensorflow.keras.preprocessing.sequence import pad_sequences

def encode_and_pad(sequences):

    # One-hot encode the sequences
    one_hot = [to_categorical(seq, num_classes=len(tokenizer.
char_index)) for seq in sequences]
```

(continued)

Listing 7.9 (continued)

```
    # Pad sequences to make them of equal length
    padded = pad_sequences(one_hot, padding='post', dtype='float32')

    # Transpose to get the desired shape: n_x, m, T_x
    transposed = np.transpose(padded, axes=(2, 0, 1))

    return transposed

def prepare_sequences(dataset, tokenizer):

    # Encode the input sequence (X)
    X_encoded = [tokenizer.encode(name) for name in dataset]

    # Shift the encoded sequence to create the output sequence (Y)
    Y_encoded = [encoded[1:] + [0] for encoded in X_encoded] # Shift
and pad with 0

    # Process the input and output sequences
    X = encode_and_pad(X_encoded)
    Y = encode_and_pad(Y_encoded)

    return X, Y
```

The "`encode_and_pad`" function takes a list of sequences as input. It first one-hot encodes the sequences, transforming each token into a binary vector of length equal to the total number of unique tokens. It then pads these one-hot encoded sequences to ensure they all have the same length, filling in any shorter sequences with zeros at the end. Finally, it transposes the padded sequences to get the desired shape for our model's input. The "`prepare_sequences`" function, on the other hand, prepares the input and output sequences for training the model. It first encodes the input sequence, converting each token into its corresponding token ID. It then creates the output sequence by shifting the encoded input sequence and padding it with a zero at the end. This is done because the task of the model is to predict the next token in the sequence, given the previous tokens. Finally, it processes these input and output sequences using the "`encode_and_pad`" function.

By using these functions, we can effectively prepare our data for training the recurrent neural network model. This will enable the model to learn the patterns in the family names of Angiosperms and generate creative new names.

Let us now run the function to generate X and Y:

```
X, Y = prepare_sequences(data, tokenizer)
```

7.6.3 Training

The next phase in the process is the training of the model. This involves updating the parameters of our model and iterating over a specified number of epochs:

Listing 7.10 Training RNN Model

```
def update_parameters(parameters, gradients, learning_rate):

    parameters['Wax'].assign_sub(learning_rate * gradients[0])
    parameters['Waa'].assign_sub(learning_rate * gradients[1])
    parameters['Wya'].assign_sub(learning_rate * gradients[2])
    parameters['ba'].assign_sub(learning_rate * gradients[3])
    parameters['by'].assign_sub(learning_rate * gradients[4])

    return parameters

def train(X, Y, parameters, num_epochs=1000, learning_rate=0.01):
    n_x, m, T_x = X.shape
    n_y, n_a = parameters["Wya"].shape

    for i in range(num_epochs):
        with tf.GradientTape() as tape:
            # Watch parameters
            tape.watch(parameters)

        # Forward propagation
        A, Y_hat = rnn_forward_propagation(X, parameters)

        # Compute loss
        loss = loss_function(Y_hat, Y)

    # Gradients through automatic differentiation
    gradients = tape.gradient(loss, [parameters['Wax'],
parameters['Waa'], parameters['Wya'], parameters['ba'],
parameters['by']])
```

(continued)

Listing 7.10 (continued)

```
    # Gradient clipping
    gradients = gradient_clipping(gradients, max_value = 5)

    # Update parameters
    parameters = update_parameters(parameters, gradients,
learning_rate)

    if i % 200 == 0:
        text = generate_text(parameters, seed=10)
        print('Epoch: %d, Loss: %f' % (i, loss))
        print (text)

return parameters
```

The "`update_parameters`" function is responsible for updating the parameters of our model. It does this by subtracting the product of the learning rate and the gradient from each parameter. This is done in accordance with the gradient descent algorithm, which is used to minimize the loss function. The "`train`" function is where the actual training of the model takes place. It begins by extracting the shape of the input data and the parameters. It then enters a loop that runs for a specified number of epochs.

Within each epoch, the function performs the following steps:

1. It watches the parameters for changes, which is necessary for automatic differentiation.
2. It performs forward propagation to compute the predicted output.
3. It computes the loss between the predicted output and the actual output.
4. It computes the gradients of the loss with respect to the parameters using automatic differentiation.
5. It applies gradient clipping to prevent the gradients from becoming too large, which can lead to numerical instability.
6. It updates the parameters using the "`update_parameters`" function.

Additionally, every 200 epochs, the function generates some text using the current parameters and prints the epoch number, the loss, and the generated text. This allows us to monitor the progress of the training and see how the generated text improves over time. By the end of the training, the function returns the updated parameters, which can then be used for generating new family names for Angiosperms.

Let us now execute the training function setting the number of nodes in the hidden state to 50. This initiates the training process and triggers the generation of creative Angiosperms family names by the RNN as it continues to learn:

```
n_x, m, T_x = X.shape
n_y = Y.shape[0]
n_a = 50 # size for the hidden state

parameters = initialize_parameters(n_x, n_a, n_y)
parameters = train(X, Y, parameters, num_epochs=2000,
learning_rate=0.1)

Epoch: 0, Loss: 2.557993
< gnzzzarlfxezncldjm
Epoch: 200, Loss: 1.229438
qiimhccaceje
Epoch: 400, Loss: 1.140713
gepoaceapaceao
Epoch: 600, Loss: 1.097014
thhirjlnaceae
Epoch: 800, Loss: 1.066846
potaeengoraciaee cea
Epoch: 1000, Loss: 1.043057
nuomadiaceae
Epoch: 1200, Loss: 1.023003
lyiiyaceae
Epoch: 1400, Loss: 1.005425
bariogapmaceae
Epoch: 1600, Loss: 0.989662
setiomaraceae
Epoch: 1800, Loss: 0.975377
bryseroaceae
```

We can see from the above output, specifically at epoch 0, the generated name appears nonsensical. However, as the training progresses, model starts to learn to generate interesting names. After 1000 epochs, the model has successfully learned to generate to the naming convention of Angiosperms family names, specifically, ending the names with the suffix "-aceae." This demonstrates the model's ability to capture and replicate the inherent patterns in the data it was trained on.

7.7 Summary

- RNNs maintain information over time through hidden states, making them suitable for tasks such as machine translation, speech recognition, and sentiment classification.

- RNN architectures include many-to-many, one-to-many, and many-to-one patterns, which are useful for tasks such as machine translation, image captioning, sentiment analysis, and creative text generation.
- Tokenization involves breaking down text into tokens and converting them into token IDs, allowing neural networks to understand and process human-readable text.
- Types of tokenization include character-level, word-level, and sub-word tokenization.
- Deep RNNs are created by stacking RNN layers for enhanced learning, while Bidirectional RNNs process sequences in both directions.
- RNN forward propagation involves computing hidden states and predictions, emphasizing the importance of temporal dependencies for sequence modeling tasks.
- Backpropagation in RNNs involves computing gradients using the chain rule and the loss function.
- Issues like vanishing and exploding gradients are mitigated using techniques like gradient clipping and specialized RNN architectures (e.g., LSTM, GRU).
- Text generation tasks involve predicting the next token in a sequence based on learned probabilities.
- Techniques like sampling with temperature control enhance creativity and diversity in generated text.

References

Bahdanau, D., Cho, K., & Bengio, Y. (2014). Neural machine translation by jointly learning to align and translate. *arXiv Preprint arXiv:1409.0473*.

Bengio, Y., Simard, P., & Frasconi, P. (1994). Learning long-term dependencies with gradient descent is difficult. *IEEE Transactions on Neural Networks*, *5*(2), 157–166.

Bishop, C. M., & Nasrabadi, N. M. (2006). *Pattern recognition and machine learning* (Vol. 4). Springer.

Chen, X., Wu, S. Z., & Hong, M. (2020). Understanding gradient clipping in private SGD: A geometric perspective. *Advances in Neural Information Processing Systems*, *33*, 13773–13782.

Cho, K., Van Merriënboer, B., Gulcehre, C., Bahdanau, D., Bougares, F., Schwenk, H., & Bengio, Y. (2014a). Learning phrase representations using RNN encoder-decoder for statistical machine translation. *arXiv Preprint arXiv:1406.1078*.

Cho, K., Van Merriënboer, B., Gulcehre, C., Bahdanau, D., Bougares, F., Schwenk, H., & Bengio, Y. (2014b). Learning phrase representations using RNN encoder-decoder for statistical machine translation. *arXiv Preprint arXiv:1406.1078*.

Collobert, R., Weston, J., Bottou, L., Karlen, M., Kavukcuoglu, K., & Kuksa, P. (2011). *Natural language processing (almost) from scratch*.

Goodfellow, I. (2016). *Deep learning*. MIT press.

Graves, A., Mohamed, A., & Hinton, G. (2013). Speech recognition with deep recurrent neural networks. *2013 IEEE International Conference on Acoustics, Speech and Signal Processing*, 6645–6649.

Gu, J., Cho, K., & Li, V. O. (2017). Trainable greedy decoding for neural machine translation. *arXiv Preprint arXiv:1702.02429*.

Hochreiter, S., & Schmidhuber, J. (1997). Long short-term memory. *Neural Computation*, *9*(8), 1735–1780.

Holtzman, A., Buys, J., Du, L., Forbes, M., & Choi, Y. (2019). The curious case of neural text degeneration. *arXiv Preprint arXiv:1904.09751*.

Jozefowicz, R., Zaremba, W., & Sutskever, I. (2015). An empirical exploration of recurrent network architectures. *International Conference on Machine Learning*, 2342–2350.

Karpathy, A., & Fei-Fei, L. (2015). Deep visual-semantic alignments for generating image descriptions. *Proceedings of the IEEE Conference on Computer Vision and Pattern Recognition*, 3128–3137.

LeCun, Y., Bottou, L., Orr, G. B., & Müller, K.-R. (2002). Efficient backprop. In *Neural networks: Tricks of the trade* (pp. 9–50). Springer.

Mikolov, T., Karafiát, M., Burget, L., Cernockŷ, J., & Khudanpur, S. (2010). Recurrent neural network based language model. *Interspeech*, *2*(3), 1045–1048.

Pang, B., Zha, K., Cao, H., Shi, C., & Lu, C. (2019). Deep RNN framework for visual sequential applications. *Proceedings of the IEEE/CVF Conference on Computer Vision and Pattern Recognition*, 423–432.

Pascanu, R., Mikolov, T., & Bengio, Y. (2013). On the difficulty of training recurrent neural networks. *International Conference on Machine Learning*, 1310–1318.

Schuster, M., & Paliwal, K. K. (1997). Bidirectional recurrent neural networks. *IEEE Transactions on Signal Processing*, *45*(11), 2673–2681.

Sennrich, R., Haddow, B., & Birch, A. (2015). Neural machine translation of rare words with subword units. *arXiv Preprint arXiv:1508.07909*.

Sutskever, I., Vinyals, O., & Le, Q. V. (2014). Sequence to sequence learning with neural networks. *Advances in Neural Information Processing Systems*, *27*.

Werbos, P. J. (1994). *The roots of backpropagation: From ordered derivatives to neural networks and political forecasting*. John Wiley & Sons.

Zhang, J., He, T., Sra, S., & Jadbabaie, A. (2019). Why gradient clipping accelerates training: A theoretical justification for adaptivity. *arXiv Preprint arXiv:1905.11881*.

Chapter 8
Crafting Long Short-Term Memory Networks

This chapter covers:

- Introducing long short-term memory (LSTM) networks and their components.
- Explaining the mechanisms of cell states and gates in LSTMs.
- Examining the process of forward propagation in LSTMs.
- Exploring the encoder–decoder architecture and its significance in machine translation.
- Detailing the training process of LSTM networks.
- Demonstrating practical applications of LSTMs in machine translation.

In the previous chapter, we covered the recurrent neural networks (RNNs) and their limitations, particularly the vanishing gradient problem. This chapter introduces long short-term memory networks (LSTMs), a neural network architecture designed to mitigate these limitations. By incorporating mechanisms such as cell states and gates, LSTMs have transformed the landscape of machine translation. We will explore the inner workings of LSTMs, their forward propagation process, and the role of the encoder–decoder architecture in modern applications. This chapter gives a detailed overview of the processes involved in training and implementing LSTMs.

8.1 LSTM Representation

In the previous chapter, we discussed recurrent neural networks (RNNs) and how they are trained for sequence modeling. However, RNNs suffer from the vanishing gradient problem due to the challenge of capturing temporal dependencies, particularly with longer sequences (Bengio et al. 1994). In order to solve the vanishing gradient problem in RNN, a new type of RNN architecture is introduced that uses special units in addition to the standard units. The new architecture is called long short-term memory (LSTM) networks (Hochreiter and Schmidhuber 1997).

T. Islam, *Hands-on Deep Learning*, https://doi.org/10.1007/978-3-032-00488-8_8

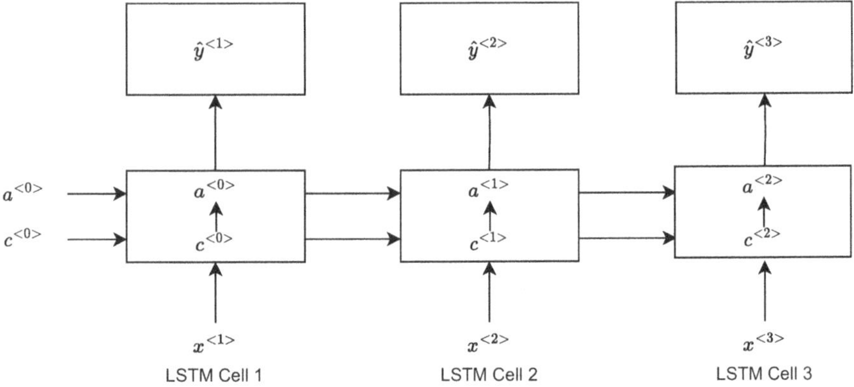

Fig. 8.1 A visual representation of the LSTM architecture in a many-to-many configuration

There are two new concepts in LSTM compared to the RNN. The first is "cell state" or "memory cell" $c^{\langle t \rangle}$, which is different from $a^{\langle t \rangle}$ as we have seen for RNNs. The addition of the "cell state" helps to keep the information in memory for longer sequences. The second concept is called "gates." The gates regulate when and how information enters the memory cell $c^{\langle t \rangle}$ and the hidden state or activations $a^{\langle t \rangle}$. Figure 8.1 provides a visual representation of the LSTM architecture in a many-to-many configuration.

We have seen previously that RNNs have a single layer that transforms the input and hidden state into an output and next hidden state. This can lead to issues such as vanishing gradients when dealing with long sequences. In contrast, LSTMs have a more complex structure with gates controlling the state updates. This allows LSTMs to better capture long-term dependencies compared to RNNs (Yu et al. 2019). The addition of the "cell state" and the "gates" in LSTMs (Gers et al. 2000) provides a more controlled way of updating and retrieving information, thereby mitigating the vanishing gradient problem as seen in RNNs. This is why LSTMs are often preferred for tasks involving longer sequences.

8.2 LSTM Forward Propagation

8.2.1 Single LSTM Cell Step

Let us now discuss on the gates, and how gates are used to update the cell sates and hidden states in a single LSTM cell step. LSTMs introduce three types of gates that help regulate the flow of information:

- Forget gate
- Update gate
- Output gate

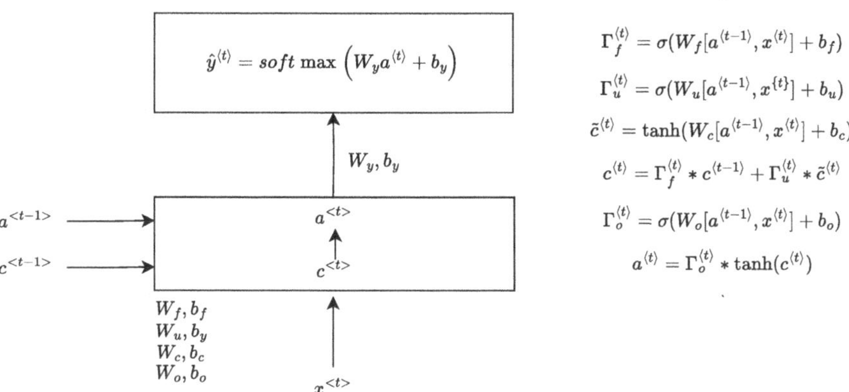

Fig. 8.2 Workings and equations of a single LSTM cell step

Figure 8.2 provides the workings and equations of a single LSTM cell step. The process begins with the formation of a candidate cell. This is done by taking the input and the previous hidden state and passing them through a function, typically a tanh activation function. The old memory cell state is then updated to the new cell state. The forget and update gates collaboratively determine this, using candidate cell. The final step involves the calculation of the hidden state using the output gate. This gate operates on the cell state, applies a hyperbolic tangent function (tanh) to constrain the values between −1 and 1, and then multiplies it by the output of a sigmoid gate. This results in the generation of the new hidden state, ready for the next time step in the sequence.

The parameters to be learned and applied in a single LSTM cell step are:

- *Weight matrices*: These are the matrices that transform the input and hidden state into the gate values. There are separate weight matrices for the forget gate (W_f), update gate (W_u), candidate cell (W_c), and output gate (W_o).
- *Bias vectors*: These are the vectors that are added to the transformed input and hidden state. There are separate bias vectors for the forget gate b_f, update gate (b_u), candidate cell (b_c), and output gate (b_o).

These parameters are learned during the training process of the LSTM network through backpropagation and gradient descent (Goodfellow 2016). The goal is to adjust these parameters in a way that minimizes the difference between the network's predictions and the actual values (the loss). Once learned, these parameters are used to perform the LSTM cell operations on new inputs.

We can initialize these parameters with He initialization (He et al. 2015) as shown in the following code:

Listing 8.1 Initialization of LSTM Parameters

```
def initialize_parameters(n_x, n_a, n_y):

    Wf = tf.Variable(tf.random.normal([n_a, n_a + n_x]) * tf.sqrt
(2./(n_a + n_x)))
    bf = tf.Variable(tf.zeros([n_a, 1]))
    Wu = tf.Variable(tf.random.normal([n_a, n_a + n_x]) * tf.sqrt
(2./(n_a + n_x)))
    bu = tf.Variable(tf.zeros([n_a, 1]))
    Wc = tf.Variable(tf.random.normal([n_a, n_a + n_x]) * tf.sqrt
(2./(n_a + n_x)))
    bc = tf.Variable(tf.zeros([n_a, 1]))
    Wo = tf.Variable(tf.random.normal([n_a, n_a + n_x]) * tf.sqrt
(2./(n_a + n_x)))
    bo = tf.Variable(tf.zeros([n_a, 1]))
    Wy = tf.Variable(tf.random.normal([n_y, n_a]) * tf.sqrt(2./
n_a))
    by = tf.Variable(tf.zeros([n_y, 1]))

    parameters = {"Wf": Wf,
                  "Wu": Wu,
                  "Wc": Wc,
                  "Wo": Wo,
                  "Wy": Wy,
                  "bf": bf,
                  "bu": bu,
                  "bc": bc,
                  "bo": bo,
                  "by": by,
                  }

    return parameters
```

Once the parameters are learnt, we can then apply these parameters for LSTM forward propagation. The single LSTM cell step calculation for forward propagation will then look like this:

Listing 8.2 Single LSTM Cell Step

```
def lstm_single_cell_step_forward(x_t, a_t_minus_1, c_t_minus_1,
parameters):
    Wf = parameters["Wf"]
    bf = parameters["bf"]
    Wu = parameters["Wu"]
    bu = parameters["bu"]
    Wc = parameters["Wc"]
    bc = parameters["bc"]
    Wo = parameters["Wo"]
    bo = parameters["bo"]
    Wy = parameters["Wy"]
    by = parameters["by"]

    concat = tf.concat([a_t_minus_1, x_t], axis=0)

    f_t = tf.sigmoid(tf.matmul(Wf, concat) + bf) # forget_gate
    u_t = tf.sigmoid(tf.matmul(Wu, concat) + bu) # update_gate
    cc_t = tf.tanh(tf.matmul(Wc, concat) + bc) # candidate_cell_state_t
    c_t = f_t * c_t_minus_1 + u_t * cc_t # cell_state
    o_t = tf.sigmoid(tf.matmul(Wo, concat) + bo)
    a_t = o_t * tf.tanh(c_t)

    y_t_hat = tf.nn.softmax(tf.matmul(Wy, a_t) + by, axis=0)

    return a_t, c_t, y_t_hat
```

The function takes as input data at a given time step, the activation (hidden state) and cell state from the previous time step, and a set of parameters. It begins by concatenating the previous activation and the current input. It then computes the forget gate, which uses a sigmoid function to decide what information to discard from the cell state. Simultaneously, the update gate determines how much of the new information should be stored. A candidate cell state is created with potential new values to be added to the state. The cell state is then updated based on the output of the forget gate, the update gate, and the candidate cell state. Finally, the output gate decides the next hidden state, which is then returned by the function along with the new cell state and the predicted output. This single step is repeated for each element in the input sequence, allowing the LSTM to maintain long-term dependencies in the data (Lipton et al. 2015). We will cover them next.

Table 8.1 The dimensions of various elements involved in an LSTM's forward propagation

Inputs, activations, outputs, and parameters	Description	Dimensions
x	Input matrix across all time steps	(n_x, m, T_x)
a	Hidden state across all time steps	(n_a, m, T_x + 1)
c	Cell state across all time steps	(n_a, m, T_x + 1)
y	Predictions across all time steps	(n_y, m, T_y)
W_f	Weight matrix of the forget gate	(n_a, n_a + n_x)
W_u	Weight matrix of the update gate	(n_a, n_a)
W_c	Weight matrix of the candidate cell	(n_y, n_a)
W_o	Weight matrix of the output gate	(n_a, n_a + n_x)
W_y	Weight matrix of the prediction	(n_y, n_a)
b_f	Bias vector of the forget gate	(n_a, n_a + n_x)
b_u	Bias vector of the update gate	(n_a, 1)
b_c	Bias vector of the candidate cell	(n_a, 1)
b_o	Bias vector of the output gate	(n_y, 1)

8.2.2 Combining Cell Steps

Just like recurrent neural networks (RNNs), the forward propagation in long short-term memory networks (LSTMs) is also a process of chaining individual LSTM cell steps together. This single LSTM cell step process is repeated for each time step in the input sequence, with the output of one LSTM cell serving as the input to the next. This chain-like nature allows the LSTM to effectively capture temporal dependencies in the input data.

Let us now go over the dimensions of the inputs, outputs, and parameters involved in the forward propagation of a long short-term memory (LSTM) network. Table 8.1 provides an overview of the dimensions of various elements involved in an LSTM's forward propagation.

Here is a code snippet that demonstrates the forward propagation of an LSTM, which integrates the computations of all individual cell steps:

Listing 8.3 LSTM Forward Propagation

```
def lstm_forward_propagation(x, parameters, a0=None, c0=None):
    n_x, m, T_x = x.shape
    n_y, n_a = parameters["Wy"].shape

    a = tf.TensorArray(dtype=tf.float32, size=T_x + 1)
    c = tf.TensorArray(dtype=tf.float32, size=T_x + 1)
    y_hat = tf.TensorArray(dtype=tf.float32, size=T_x)
```

(continued)

Listing 8.3 (continued)

```
    if a0 is None:
        a0 = tf.zeros([n_a, m])

    if c0 is None:
        c0 = tf.zeros([n_a, m])

    a = a.write(0, a0)
    c = c.write(0, c0)

    for t in tf.range(1, T_x + 1):
        a_t_minus_1 = a.read(t - 1)
        c_t_minus_1 = c.read(t - 1)
        a_t, c_t, y_t_hat = lstm_single_cell_step_forward(x[:, :,
t - 1], a_t_minus_1, c_t_minus_1, parameters)
        a = a.write(t, a_t)
        c = c.write(t, c_t)
        y_hat = y_hat.write(t - 1, y_t_hat)

    # Ensure that the final tensors have the correct shapes
    a = a.stack()
    c = c.stack()
    y_hat = y_hat.stack()

    a = tf.transpose(a, [1, 2, 0])
    c = tf.transpose(c, [1, 2, 0])
    y_hat = tf.transpose(y_hat, [1, 2, 0])

    return a, c, y_hat
```

8.3 Encoder–Decoder LSTM

The encoder–Decoder architecture (Sutskever et al. 2014) with long short-term memory (LSTM) units is a popular choice for various tasks in natural language processing (NLP) and other fields. For instance, in machine translation, the encoder processes the input sentence and compresses the information into a context vector, and the decoder then generates the translated sentence from this context vector. Similarly, encoder–decoder LSTM can be used to generate a shorter version of a given text, maintaining the key information and overall meaning. In speech recognition (Graves et al. 2013a, b), the encoder processes the audio signal as input and the decoder generates the corresponding text. Encoder–decoder LSTM can also be

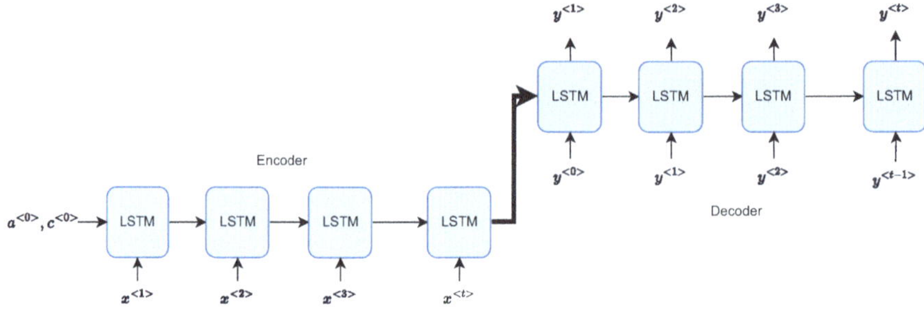

Fig. 8.3 An illustration of encoder–decoder LSTM architecture

used in creating conversational AI chatbots (Young et al. 2018) where the encoder processes the input sentence and the decoder generates the response.

Figure 8.3 illustrates the encoder–decoder architecture. Let us consider a practical example to elucidate this concept. Suppose we have a sentence in French: "*Paris est la capitale de la France*," which translates to "*Paris is the capital of France*" in English. In this scenario, the encoder begins by processing the French sentence. It sequentially reads each word, capturing the semantic essence of the sentence and condensing this information into a context vector (Bahdanau et al. 2014; Cho et al. 2014), often referred to as the final state of the encoder. This context vector, a compact representation of the entire French sentence, is then passed on to the decoder. The decoder's task is to generate the English translation. It is typically provided with a start token as its initial input, which signals the beginning of the output sequence. After that, the decoder generates the next token, which could be a word in the case of machine translation. This generated token is then fed back into the decoder as input, and the process continues until a special end token is generated, signaling the completion of the output sequence. Note that we use a process referred to as teacher forcing (Williams and Zipser 1989) during training, where the actual output (instead of the generated output) is used as the next input to the decoder. This approach helps the model learn more effectively.

8.4 Building an End-to-End LSTM Model

8.4.1 Machine Translation Problem

This time we will be implementing LSTM for machine translation. Specifically, we will be focusing on the translation of numerical values from French to English, expressed in words. The numbers are expressed in words, and the model must understand the linguistic rules that govern how numbers are formed and expressed in both French and English.

Here is an example to illustrate what we aim to achieve:

Input in French: “*huit cent vingt-quatre*”.
Translation in English: “*eight hundred and twenty-four*”.

Let us first create the dataset using `num2words` library. The following lines of code show how to create such dataset using `num2words` library:

Listing 8.4 Creating Machine Translation Data

```
from num2words import num2words

def create_french_to_english_data():
    start = 1
    end = 1000

    # Create the dataset
    source = []
    target = []
    for i in range(start, end+1):
        # Convert the number to words in French and English
        words = '<start> ' + num2words(i, lang='fr') + ' <end>'
        source.append(words)
        words = '<start> ' + num2words(i, lang='en') + ' <end>'
        target.append(words)

    return source, target
```

Let us now generate source and target sentences by executing the above function:

```
source, target = create_french_to_english_data()
```

We will now split the data into train and test set (Pedregosa et al. 2011):

Listing 8.5 Function to Split the Source and Target Data into Train and Test Set

```
from sklearn.model_selection import train_test_split

def split_data(source, target):

    train_source, test_source, train_target, test_target =
train_test_split(
        source, target, test_size=0.2, random_state=42
    )

    return train_source, test_source, train_target, test_target
```

We will execute this as:

```
train_source, test_source, train_target, test_target = split_data
(source, target)
```

8.4.2 Sequence Preparation

We know that when working with language data, we need to prepare our data in a specific format. This process involves several steps:

1. *Tokenization*: This is the process of converting words into numerical representations, also known as token IDs (Grefenstette 1999). Each unique word in our data is assigned a unique integer ID.
2. *Padding*: Since sequences of words in language data can have varying lengths, we need to ensure that all sequences have the same length to feed them into our model. We do this by either truncating longer sequences or padding shorter ones with zeros (Chollet 2021; Goodfellow 2016).
3. *One-hot encoding*: After padding, we convert these sequences of token IDs into a one-hot representation. In a one-hot encoded vector, each word is represented as a vector where all elements are 0, except for the position corresponding to the word's token ID, which is set to 1.

In the following code, we prepare sequences for both the source language data and the target language data:

Listing 8.6 Encoding, Padding, and Preparing Sequences

```
from tensorflow.keras.utils import to_categorical
from tensorflow.keras.preprocessing.sequence import
pad_sequences

def encode_and_pad(sequences):

    padded = pad_sequences(sequences, maxlen=8, padding='post',
truncating='post', dtype='float32')
    one_hot = to_categorical(padded, num_classes=len(tokenizer.
word_index)+1)
    transposed = np.transpose(one_hot, axes=(2, 0, 1))
    return transposed

def prepare_sequences(source, target, tokenizer):
```

(continued)

Listing 8.6 (continued)
```
    X_encoder = tokenizer.texts_to_sequences(source)

    Y_encoded = tokenizer.texts_to_sequences(target)
    Y_decoder_input = Y_encoded
    Y_decoder_output = [encoded[1:] + [0] for encoded in Y_encoded]

    X_encoder = encode_and_pad(X_encoder)
    Y_decoder_input = encode_and_pad(Y_decoder_input)
    Y_decoder_output = encode_and_pad(Y_decoder_output)

    return X_encoder, Y_decoder_input, Y_decoder_output
```

For the target language data, we prepare two versions of sequences: one for input to the decoder part of our model and one for output. The output sequences are just the input sequences shifted by one time step, with a zero appended at the end to maintain the same length. This way, our LSTM model can be trained to predict the next word in the sequence given the current and previous words (teacher forcing). This is how we prepare our sequences for training a language model such as LSTM:

```
X_train_encoder, Y_train_decoder_input, Y_train_decoder_output =
prepare_sequences(train_source, train_target, tokenizer)
X_test_encoder, Y_test_decoder_input, Y_test_decoder_output =
prepare_sequences(test_source, test_target, tokenizer)
```

8.4.3 Training

We now work on writing our training code. We first define the loss function, which is averaged over time steps as well as examples:

Listing 8.7 Loss Function over Time Steps
```
def loss_function(y_hat, y):

    # Compute loss at each timestep for each example (for n_classes/
one hot)
    loss = -tf.reduce_sum(y * tf.math.log(y_hat), axis=0)

    # Average over timesteps for each example
    loss = tf.reduce_mean(loss, axis=-1)
```

(continued)

Listing 8.7 (continued)

```
    # Average over examples
    loss = tf.reduce_mean(loss)

    return loss
```

Next is coding up the training function:

Listing 8.8 Training LSTM Model

```
def   train(X_encoder,   Y_decoder_input,   Y_decoder_output,
X_test_encoder, Y_test_decoder_input, Y_test_decoder_output,
parameters_encoder, parameters_decoder, num_epochs=1000):

    optimizer = tf.keras.optimizers.legacy.Adam()

    n_x, m, T_x = X_encoder.shape
    n_y, n_a = parameters_decoder["Wy"].shape

    for i in range(num_epochs):
        with tf.GradientTape(persistent=True) as tape:
            tape.watch(parameters_encoder)
            tape.watch(parameters_decoder)

            # Forward propagation through the encoder
            A, C, _ = lstm_forward_propagation(X_encoder,
parameters_encoder)

            # Use the final hidden state of the encoder as the initial
hidden state of the decoder
            A0 = A[:, :, -1]
            C0 = C[:, :, -1]

            # Forward propagation through the decoder
            _, _, Y_hat = lstm_forward_propagation(Y_decoder_input,
parameters_decoder, A0, C0)

            # Compute loss
            loss = loss_function(Y_hat, Y_decoder_output)
```

(continued)

Listing 8.8 (continued)

```
        # Compute gradients
        gradients_encoder = tape.gradient(loss, parameters_encoder)
        gradients_decoder = tape.gradient(loss, parameters_decoder)

        optimizer.apply_gradients(zip(gradients_encoder.values(),
parameters_encoder.values()))
        optimizer.apply_gradients(zip(gradients_decoder.values(),
parameters_decoder.values()))

        if i % 50 == 0:
            print('Epoch: %d, Loss: %f' % (i, loss))

    return parameters_encoder, parameters_decoder
```

In each epoch, the first step is to perform forward propagation through the encoder and decoder. The encoder processes the input data and the final hidden state of the encoder is used as the initial hidden state of the decoder. The decoder then generates the output data. After the forward propagation, we compute the loss. The loss function measures how far the model's predictions are from the actual target. The goal of training is to minimize this loss. Once the loss is computed, we perform backward propagation to compute the gradients of the loss with respect to the model's parameters. These gradients indicate the direction in which we need to adjust our parameters to reduce the loss. The optimizer uses these gradients to update the model's parameters. This update step is what allows the model to learn from the data.

Let us now train for 500 epochs using 512 LSTM units:

```
n_x, m, T_x = X_train_encoder.shape
n_y = Y_train_decoder_input.shape[0]
n_a = 512 # Nodes for the hidden state

parameters_encoder = initialize_parameters(n_x, n_a, n_y)
parameters_decoder = initialize_parameters(n_x, n_a, n_y)
parameters = train(X_train_encoder, Y_train_decoder_input,
Y_train_decoder_output, X_test_encoder, Y_test_decoder_input,
Y_test_decoder_output, parameters_encoder, parameters_decoder,
num_epochs=500)
```

```
Epoch: 0, Loss: 4.075758
Epoch: 50, Loss: 0.881784
Epoch: 100, Loss: 0.721993
```

```
Epoch: 150, Loss: 0.413168
Epoch: 200, Loss: 0.215170
Epoch: 250, Loss: 0.022495
Epoch: 300, Loss: 0.008216
Epoch: 350, Loss: 0.004773
Epoch: 400, Loss: 0.003286
Epoch: 450, Loss: 0.002394
```

We can see that loss has been decreasing steadily. We already know this, but let us reemphasize. When we say "the loss has been decreasing steadily," we are indicating that with each iteration or epoch of the training process, the model is getting better at predicting the output for the given inputs. This is a positive sign, implying that the model is effectively learning from the data. It is adjusting its internal parameters in a way that reduces the discrepancy between its predictions and the actual outcomes. The steady decrease in loss implies a consistent improvement in the model's performance, albeit on training data.

8.4.4 Prediction

We also know that prediction is all about forward propagation with learned parameters. The following lines of code show how to predict the probability of each possible output in the sequence:

Listing 8.9 LSTM Prediction

```
def predict(X_encoder, Y_decoder_input, parameters_encoder,
parameters_decoder):
    # Encoder
    A, C, _ = lstm_forward_propagation(X_encoder,
parameters_encoder)

    # Use the final hidden state of the encoder as the initial hidden
state of the decoder
    A0 = A[:, :, -1]
    C0 = C[:, :, -1]

    # Forward propagation through the decoder
    _, _, Y_hat = lstm_forward_propagation(Y_decoder_input,
parameters_decoder, A0, C0)

    return Y_hat
```

The process begins with the encoder part of the model processing the input data. This results in a set of hidden states and cell states. The final hidden state and cell state of the encoder are then used as the initial states for the decoder, effectively providing the decoder with a "context" or "memory" of what the encoder has processed. The decoder then processes its own input data, starting from these initial states, and generates a set of predictions.

For the sake of reference, the following is the code to predict and calculate the perplexity score on the test set:

```
Y_hat = predict(X_test_encoder, Y_test_decoder_input,
parameters_encoder, parameters_decoder)
loss = loss_function(Y_hat, Y_test_decoder_output)
perplexity = tf.exp(loss).numpy()
print (perplexity)
```

```
1.0135947
```

8.4.5 Generate Translation

The prediction process outlined above employs the teacher forcing technique (Williams and Zipser 1989). In this method, we use the actual output sequences (the ground truth that is shifted) as input to the decoder during training. This approach helps the model learn more effectively as it has access to the full context of the input sequence.

However, in a real-world scenario, we will not have access to the actual output sequence while making predictions. Instead, we generate each word sequentially, using the model's own predictions as input for the next step. Here is the code, which demonstrates this process:

Listing 8.10 Generating Translation

```
def translate(text):
    encoder_input = tokenizer.texts_to_sequences([text])
    encoder_input = encode_and_pad(encoder_input)

    sequences = []
    word = '<start>'
    idx = tokenizer.word_index[word]
    sequences.append(idx)

    translated_text = ''
    for t in range(T_y):
```

(continued)

Listing 8.10 (continued)

```
        translated_text += word + ' '
        decoder_input = encode_and_pad([sequences])
        y_hat = predict(encoder_input, decoder_input, 
parameters_encoder, parameters_decoder)
        idx = np.argmax(y_hat[:, 0, t])
        sequences.append(idx)
        word = tokenizer.index_word[idx]
        if word == '<end>':
            translated_text += word + ' '
            break

    return translated_text
```

In this function, text is the input sentence to be translated. The function first converts the text into sequences and pads them. It then initializes the sequences with the start token. For each time step, it adds the current word to the translated text, prepares the decoder input, predicts the next word, and adds the predicted word to the sequences. The process continues until the end token is found. The function then returns the translated text. This approach allows the model to generate translations one word at a time, using its own predictions as input for subsequent steps.

The following code generates few example translations from the test set:

Listing 8.11 Generating Example Translations

```
def generate_example_translations():
    indices = np.random.choice(200, 5, replace=False)
    for i in range(5):
        text = test_source[indices[i]]
        translated_text = translate(text)
        print ("----------------")
        print ("French text:", test_source[indices[i]])
        print ("Translation:", translated_text)
        print ("Ground truth:", test_target[indices[i]])

    return None
```

Let us generate few example translations:

```
generate_example_translations()
```

```
----------------
French text: <start> cinq cent quatre-vingt-quatre <end>
Translation: <start> five hundred and eighty four <end>
Ground truth: <start> five hundred and eighty-four <end>
----------------
French text: <start> six cent trente-six <end>
Translation: <start> six hundred and thirty six <end>
Ground truth: <start> six hundred and thirty-six <end>
----------------
French text: <start> huit cent vingt-sept <end>
Translation: <start> eight hundred and twenty seven <end>
Ground truth: <start> eight hundred and twenty-seven <end>
----------------
French text: <start> cinq cent vingt-trois <end>
Translation: <start> five hundred and twenty three <end>
Ground truth: <start> five hundred and twenty-three <end>
----------------
French text: <start> huit cent vingt-quatre <end>
Translation: <start> eight hundred and twenty four <end>
Ground truth: <start> eight hundred and twenty-four <end>
```

In the produced output, we can see that our trained LSTM model is able to translate French language into English correctly.

8.5 Summary

- LSTMs (long short-term memory networks) are introduced to address the vanishing gradient problem in RNNs by incorporating "cell state" and "gates."
- LSTMs use three types of gates: forget gate, update gate, and output gate to regulate the flow of information.
- The "cell state" in LSTMs allows the network to retain information over long sequences.
- Forward propagation in LSTMs involves chaining individual LSTM cell steps together to capture temporal dependencies.
- Encoder–decoder LSTM architecture is effective for tasks like machine translation, summarization, and speech recognition.
- The training process of LSTMs involves backpropagation and gradient descent to adjust parameters, aiming to minimize the loss.
- Implementing LSTMs for machine translation requires sequence preparation, tokenization, padding, and one-hot encoding.

References

Bahdanau, D., Cho, K., & Bengio, Y. (2014). Neural machine translation by jointly learning to align and translate. *arXiv Preprint arXiv:1409.0473*.

Bengio, Y., Simard, P., & Frasconi, P. (1994). Learning long-term dependencies with gradient descent is difficult. *IEEE Transactions on Neural Networks*, *5*(2), 157–166.

Cho, K., Van Merriënboer, B., Gulcehre, C., Bahdanau, D., Bougares, F., Schwenk, H., & Bengio, Y. (2014). Learning phrase representations using RNN encoder-decoder for statistical machine translation. *arXiv Preprint arXiv:1406.1078*.

Chollet, F. (2021). *Deep learning with Python*. simon and schuster.

Gers, F. A., Schmidhuber, J., & Cummins, F. (2000). Learning to forget: Continual prediction with LSTM. *Neural Computation*, *12*(10), 2451–2471.

Goodfellow, I. (2016). *Deep learning*. MIT press.

Graves, A., Jaitly, N., & Mohamed, A. (2013a). Hybrid speech recognition with deep bidirectional LSTM. *2013 IEEE Workshop on Automatic Speech Recognition and Understanding*, 273–278.

Graves, A., Mohamed, A., & Hinton, G. (2013b). Speech recognition with deep recurrent neural networks. *2013 IEEE International Conference on Acoustics, Speech and Signal Processing*, 6645–6649.

Grefenstette, G. (1999). Tokenization. In *Syntactic wordclass tagging* (pp. 117–133). Springer.

He, K., Zhang, X., Ren, S., & Sun, J. (2015). Delving deep into rectifiers: Surpassing human-level performance on imagenet classification. *Proceedings of the IEEE International Conference on Computer Vision*, 1026–1034.

Hochreiter, S., & Schmidhuber, J. (1997). Long short-term memory. *Neural Computation*, *9*(8), 1735–1780.

Lipton, Z. C., Berkowitz, J., & Elkan, C. (2015). A critical review of recurrent neural networks for sequence learning. *arXiv Preprint arXiv:1506.00019*.

Pedregosa, F., Varoquaux, G., Gramfort, A., Michel, V., Thirion, B., Grisel, O., Blondel, M., Prettenhofer, P., Weiss, R., Dubourg, V., & others. (2011). Scikit-learn: Machine learning in Python. *The Journal of Machine Learning Research*, *12*, 2825–2830.

Sutskever, I., Vinyals, O., & Le, Q. V. (2014). Sequence to sequence learning with neural networks. *Advances in Neural Information Processing Systems*, *27*.

Williams, R. J., & Zipser, D. (1989). A learning algorithm for continually running fully recurrent neural networks. *Neural Computation*, *1*(2), 270–280.

Young, T., Hazarika, D., Poria, S., & Cambria, E. (2018). Recent trends in deep learning based natural language processing. *IEEE Computational Intelligence Magazine*, *13*(3), 55–75.

Yu, Y., Si, X., Hu, C., & Zhang, J. (2019). A review of recurrent neural networks: LSTM cells and network architectures. *Neural Computation*, *31*(7), 1235–1270.

Chapter 9
Using Embeddings in Language Models

This chapter covers:

- Introducing the concept of embeddings and embedding layer.
- Exploring different types of embeddings and their applications.
- Analyzing how embeddings capture semantic relationships between texts.
- Learning embeddings in deep learning models.
- Providing practical examples for using embeddings in a topic classification task.

This chapter introduces embeddings, a powerful way to represent words as vectors in high-dimensional space. Embeddings help us understand the relationships between words and their contexts much better. We will explore how these embeddings work, from basic concepts to advanced techniques, and show how they capture semantic similarity between texts. Through clear examples and practical guides, we will see how embeddings are trained and used in natural language processing models.

9.1 Embedding Representation

In the past, we utilized one-hot encoding to represent tokens or words. This method is adequate for certain tasks. However, it presents a challenge for the model to understand the relationships between words. This is primarily due to the fact that the inner product of two one-hot encoded vectors does not provide any meaningful information about the relationship between the corresponding words. In other words, the one-hot encoding method treats each word as an isolated entity, without considering its context or its semantic and syntactic relationships with other words. This is a significant limitation when we are trying to capture the nuances of natural language in our models. Therefore, a more sophisticated representation is needed to capture these relationships and provide a richer representation of the text data.

T. Islam, *Hands-on Deep Learning*, https://doi.org/10.1007/978-3-032-00488-8_9

Word embeddings is one such representation. Word embeddings map words in a high-dimensional space where the location and distance between words indicate their semantic similarity (Mikolov et al. 2013a). This allows the model to understand and leverage the relationships between different words, thereby improving its performance on various natural language processing tasks.

There are several advantages of using an embedding representation over a raw one-hot encoded representation of words. First, an embedding representation tends to generalize better. This is because it captures more nuanced information about words, including their context within a sentence or document, which can lead to more accurate predictions. Second, embedding representations possess the property of analogy (Mikolov et al. 2013c). This means they can capture semantic relationships between words through a similarity metric. The words that are semantically similar will have similar vector representations in the embedding space. This allows the model to understand and leverage these relationships, thereby enhancing its ability to process and understand natural language.

9.2 Embedding Layer

The embedding representation is learned and retrieved through an embedding layer. Embedding layer converts tokenized input data into a representation of dense vectors of fixed size (Chollet 2021). The key idea is to represent tokenized categorical data as vectors in a high-dimensional space, where the distance and direction between vectors capture some semantic relationships between the words (tokens).

The embedding mechanism is depicted in Fig. 9.1. First, the raw text is converted to tokens in the form of IDs through a tokenizer. Then, each token is converted to an embedding vector through an embedding matrix.

More specifically, token embeddings are computed using a lookup table that stores the token embedding weights matrix W_E (Chollet 2021; Goodfellow 2016). This matrix, with dimensions corresponding to the vocabulary size and the embedding size, is a set of learnable parameters and is learned during training. To retrieve the token embeddings, we use token indices to access the appropriate entries in the embedding weight matrix. The process takes an integer as input and returns the corresponding token embedding as output. The equation for token embedding computation can be written as:

$$E(i) = W_E[i]$$

Here, $E(i)$ is the embedding for the input token ID i. The operation $W_E[i]$ is a lookup on the weight matrix W_E using the index i. The weight matrix W_E in the embedding layer has dimensions of $V * D$, where:

- V is the size of the vocabulary, i.e., the total number of unique tokens in the tokenizer. Each row in the matrix corresponds to a token in the vocabulary.

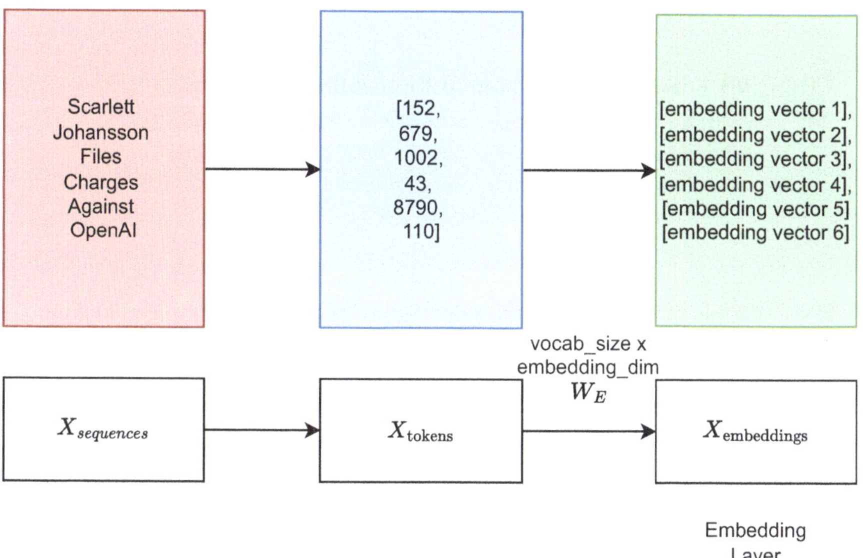

Fig. 9.1 The embedding mechanism

- D is the dimensionality of the embeddings, i.e., the size of the vector space in which tokens will be embedded. This is a hyperparameter.

This means, if we have a vocabulary of 30,000 tokens and we want to embed each token into a 256-dimensional vector space, the weight matrix W_E will then have dimensions of $30{,}000 \times 256$. Note that these embeddings are learned during the training process. The embedding layer is all about learning embedding weight matrix W_E.

Here is a code snippet that shows how to calculate embeddings given embedding matrix:

```
batch_size = 32
sequence_length = 50
vocab_size = 10000
embedding_dim = 256
W_e = np.random.randn(vocab_size, embedding_dim)

X_token_ids = tf.random.uniform([batch_size, sequence_length],
minval=0, maxval=vocab_size, dtype=tf.int32).numpy()
X_one_hot = tf.one_hot(X_token_ids, depth=vocab_size).numpy()
X_embeddings = np.dot(X_one_hot, W_e)
print (X_embeddings.mean())
```

```
-0.0015396117114564 72
```

This is essentially forward propagation for the embedding layer. We can also achieve this using `tf.nn.embedding_lookup`:

Listing 9.1 Forward Propagation of Embedding Layer

```
def embedding_forward_propagation(X_ids, W_e):
    X_embeddings = tf.nn.embedding_lookup(W_e, X_ids)
    X_embeddings = tf.transpose(X_embeddings, [2, 0, 1])
    return X_embeddings
```

If we run this, it will produce the same output as the vanilla implementation:

```
X_embeddings = embedding_forward_propagation(X_token_ids, W_e)
print (X_embeddings.numpy().mean())
```

```
-0.001539611711456472
```

9.3 Learning Embeddings

9.3.1 Embedding Models

Now the question is how can we model a neural network to learn the embeddings? Learning embeddings can be broadly divided into two types:

- *Global embedding models*: These models generate a single word embedding representation for each word in the vocabulary. For example, consider two sentences. "*Kendall Jenner is the highest paid model in the USA*" (Sentence 1) and "*Regnet is a neural network model used by Tesla*" (Sentence 2). Despite having different contexts in the two sentences, global embedding models would generate the same representation for the word "*model.*" Word2Vec, GloVe, and FastText are examples of global embedding models (Bojanowski et al. 2017; Mikolov et al. 2013a, b; Pennington et al. 2014).
- *Contextual embedding models*: These models generate a representation of each word that is based on the other words in the sentence. For example, in the above two sentences, the word "model" has different meanings based on context. Therefore, the same word "model" will have two different embeddings. BERT, ELMo, UseNet, and GPT are examples of contextual models (Cer et al. 2018; Devlin et al. 2019; Peters et al. 2018; Radford et al. 2018).

The learned embeddings can be used for transfer learning or some other tasks such as recommendation system. We will go through an example of this later.

9.3.2 Contextual Embeddings with RNN

There exists a multitude of methodologies for training a contextual embedding model. The foundational sequence-to-sequence model can be based on various architectures, such as recurrent neural networks (RNN), long short-term memory (LSTM), or the transformer architecture (Bahdanau et al. 2014; Sutskever et al. 2014; Vaswani et al. 2017). The design of the training process can also be tailored in several ways. For example, one approach can be masking certain words within a corpus and subsequently training the model to predict these masked words (Liu et al. 2019). Alternatively, we could construct a contextual model that predicts the subsequent word in a sentence. Another possibility is to develop a contextual model through a task-specific problem, such as text classification using a sequence model.

Let us look at an example of a model that utilizes contextual embeddings. As illustrated in Fig. 9.2, this model employs a many-to-one RNN architecture with embeddings. A good example of this design is the classification of news topics, which we will explore in greater detail later in this chapter. The task involves the classification of news topics according to the textual content of news stories. The primary objective of the model is to learn to classify these topics. However, during this learning process, the model also acquires the ability to represent words in a high-dimensional space, effectively learning embeddings. This means that words appearing in similar contexts, or in this case, similar types of news articles, are positioned closer to each other in this space. These learned embeddings are indeed contextual, as they encapsulate the meaning of a word within the specific context of a news article.

In essence, we previously used a one-hot encoded representation for each word at time step "*t*" and fed it into the RNN. Now, we will feed the embedding vector into the RNN instead. This change allows the model to operate on more nuanced and contextually rich representations of words, enhancing its performance in tasks such as news topic classification.

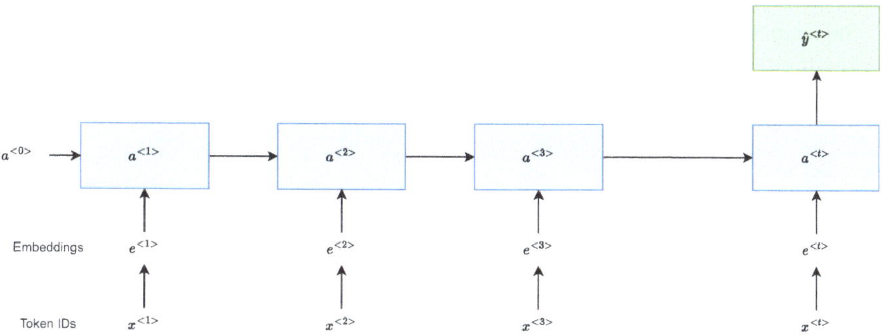

Fig. 9.2 Illustration of embeddings in a many-to-one architecture

Therefore, since we are introducing embedding layer in the network, we will have to initialize embedding weight matrix as well as RNN parameters. The initialization will look like this in the following code:

Listing 9.2 Initialization of Embedding Layer and RNN Parameters

```
def initialize_parameters(n_x, n_a, n_y, n_e, initialization="he"):
    tf.random.set_seed(210)
    We = tf.Variable(tf.random.normal([n_x, n_e]), trainable=True)

    if initialization == "he":
        Wax = tf.Variable(tf.random.normal([n_a, n_e]) * tf.sqrt
(2. / n_e), trainable=True)
        Waa = tf.Variable(tf.random.normal([n_a, n_a]) * tf.sqrt
(2. / n_a), trainable=True)
        Wya = tf.Variable(tf.random.normal([n_y, n_a]) * tf.sqrt
(2. / n_a), trainable=True)
    elif initialization == "xavier":
        print("Not Implemented")

    ba = tf.Variable(tf.zeros([n_a, 1]), trainable=True)
    by = tf.Variable(tf.zeros([n_y, 1]), trainable=True)

    parameters = {"We": We,
                  "Wax": Wax,
                  "Waa": Waa,
                  "Wya": Wya,
                  "ba": ba,
                  "by": by}

    return parameters
```

Subsequently, the forward propagation steps will be as follows. Here, we are borrowing the `rnn_forward_propagation` function from the RNN chapter:

```
X_embeddings = embedding_forward_propagation(X_token_ids,
parameters['We'])
A, Y_hat = rnn_forward_propagation(X_embeddings, parameters)
```

9.3.3 Global Embeddings with Word2Vec

Consider the sentence, "*Kendall Jenner is the highest paid model in the USA.*" If we select "paid" as our target word and set a window size of 2, the context words within this window are "the," "highest," "model," and "in." In relation to our target word "paid," these context words are labeled as positive examples. On the other hand, negative examples, which are not contextually related to the target word, are chosen randomly from the vocabulary. This forms the basis of the Skip-gram model in Word2Vec (Mikolov et al. 2013a), where the aim is to correctly classify context words (positive examples) and non-context words (negative examples) given a target word.

Here is the example code to generate skip-gram pairs and labels:

Listing 9.3 Generating Skip-Gram Word Pairs

```
import tensorflow as tf
from tensorflow.keras.preprocessing.text import Tokenizer
from tensorflow.keras.preprocessing.sequence import skipgrams

def generate_skipgram_word_pairs(text_data):

    # Tokenize the text data
    tokenizer = Tokenizer()
    tokenizer.fit_on_texts(text_data)

    # Convert text to sequences
    sequences = tokenizer.texts_to_sequences(text_data)

    # Prepare training data for skip-gram
    word_pairs, labels = [], []
    for sequence in sequences:
        pairs, lbls = skipgrams(sequence, len(tokenizer.
word_index) + 1, window_size=2)
        word_pairs.extend(pairs)
        labels.extend(lbls)

    return word_pairs, labels, tokenizer
```

Once the pairs are generated, the Skip-gram model is then trained on these pairs as a binary classification problem:

Listing 9.4 Word2Vec Model

```
class Word2Vec(tf.keras.Model):
    def __init__(self, vocab_size, embedding_dim):
        super(Word2Vec, self).__init__()
        self.target_embedding = tf.keras.layers.Embedding
(vocab_size, embedding_dim)
        self.context_embedding = tf.keras.layers.Embedding
(vocab_size, embedding_dim)

    def call(self, pair):
        target, context = pair[:, 0], pair[:, 1]
        target_emb = self.target_embedding(target)
        context_emb = self.context_embedding(context)
        dot_product = tf.reduce_sum(target_emb * context_emb,
axis=1)
        return tf.sigmoid(dot_product)
```

In this code, given a pair of words (a target word and a context word), it computes their embeddings and then calculates the dot product of these two vectors. The dot product serves as an indicator of how similar two vectors are—the greater the dot product, the more alike the vectors become. The dot product is then passed through a sigmoid function. The sigmoid function maps the potentially large positive or negative values of the dot product into a probability between 0 and 1. This is useful because we are essentially treating the Word2Vec task as a binary classification problem—given a pair of words, the model needs to learn whether it is a valid (positive) pair of target and context word or not (negative). The model can then be trained to maximize this probability for actual pairs of words that appear in the text (positive examples) and minimize it for random pairs of words (negative examples, hence the term "Negative Sampling").

9.4 Cosine Similarity

Once we have trained our embeddings, the next question is: how do we utilize them? Specifically, how can we understand the semantic similarity between two words through embeddings? The answer lies in measuring the degree of similarity between two embedding vectors. This can be achieved through a mathematical concept known as cosine similarity. By comparing the cosine of the angle between two embedding vectors, we can quantify their semantic similarity.

Cosine similarity is a way to determine how similar two nonzero vectors are in an inner product space (Singhal et al. 2001). The definition refers to the cosine of the angle that separates two vectors. This is computed by taking the dot product of the vectors and dividing it by the product of the Euclidean (L2) norms of the two vectors. The mathematical formula for cosine similarity is:

$$\text{cosine\_similarity}(a, b) = \frac{a \cdot b}{||a||_2 \times ||b||_2}$$

where:

- $a \cdot b$ is the dot product of vectors a and b,
- $||a||_2$ is the Euclidean (L2) norm of vector a,
- $||b||_2$ is the Euclidean (L2) norm of vector b.

Here is the code snippet to implement the computation of cosine similarity between two vectors:

Listing 9.5 Computing Cosine Similarity

```
import numpy as np

def cosine_similarity(vec1, vec2):
    # Compute the dot product of vec1 and vec2
    dot_product = np.dot(vec1, vec2)

    # Compute the L2 norm of vec1 and vec2
    norm_vec1 = np.sqrt(np.sum(np.square(vec1)))
    norm_vec2 = np.sqrt(np.sum(np.square(vec2)))

    # Compute the cosine similarity
    cosine_sim = dot_product / (norm_vec1 * norm_vec2)

    return cosine_sim
```

The range of cosine similarity is between -1 and 1. A value of 1 implies that the vectors are identical, 0 implies that the vectors are orthogonal (not similar), and -1 implies that the vectors are diametrically opposed (opposite). In the context of text analysis, cosine similarity is often used to measure the semantic similarity of documents, as it takes into account the orientation (but not the magnitude) of the vectors in the vector space. This makes it particularly suited for high-dimensional, sparse vectors such as those used in text representation.

9.5 Building an End-to-End Contextual Embeddings Model

9.5.1 News Topic Classification Problem

In this hands-on case study, we will be utilizing the "ag_news_subset" dataset from TensorFlow. The AG dataset is a comprehensive collection of over a million news articles (Zhang et al. 2015). Our objective is to categorize these articles into one of four distinct topics or classes. These classes are represented as 0 (World), 1 (Sports), 2 (Business), and 3 (Sci/Tech). By training a model to classify these articles into their respective classes using an embedding layer, we will be demonstrating that the model can also learn the semantic representation of the articles through these embeddings.

The embeddings we learn can be subsequently employed for news recommendation, effectively transforming this case study into a recommender system. For example, consider a recommendation system scenario. When a reader is engaged with a news article, we can leverage the power of these embeddings to suggest a different news article that is semantically similar to the one the user is currently reading. We will see that this recommendation will be made possible through the similarity of the embeddings, which capture the underlying semantic structure of the articles.

The content of the dataset looks like this:

Description	Label	Title
Witnesses in the trial of a US soldier charged with abusing prisoners at Abu Ghraib have told the court that the CIA sometimes directed abuse and orders were received from military command to toughen interrogations	0 (World)	Witness says CIA oversaw abuse at Abu Ghraib

First things first. We start by loading the data from TensorFlow dataset. We will be combining the title and description together. The code to load this dataset is shown as follows:

Listing 9.6 Loading AgNews data

```
def load_ag_news_data():

    train_dataset, dataset_info = tfds.load('ag_news_subset',
with_info=True, split='train')

    X_train = []
    y_train = []

    # Iterate over the dataset
    for example in train_dataset:
```

(continued)

Listing 9.6 (continued)

```
        train_content = 'Title: ' + example['title'].numpy().
decode('utf-8') + ' Description: ' + example['description'].numpy
().decode('utf-8')
        X_train.append(train_content)
        y_train.append(example['label'].numpy())

    test_dataset, dataset_info = tfds.load('ag_news_subset',
with_info=True, split='test')

    X_test = []
    y_test = []
    for example in test_dataset:
        test_content = 'Title: ' + example['title'].numpy().decode
('utf-8') + ' Description: ' + example['description'].numpy().
decode('utf-8')
        X_test.append(test_content)
        y_test.append(example['label'].numpy())

    return X_train, np.array(y_train), X_test, np.array(y_test)
```

We can call the above function to create our train and test set:

```
X_train, y_train, X_test, y_test = load_ag_news_data()
```

We will then apply tokenization to the raw text to convert to token IDs to feed into an embedding layer. We will also be applying padding. Here is the code snippet for tokenization:

Listing 9.7 Tokenization

```
from tensorflow.keras.preprocessing.text import Tokenizer

def tokenization(X_train, X_test, sequence_length):
    tokenizer = Tokenizer(oov_token="<OOV>")
    tokenizer.fit_on_texts(X_train)

    X_train_token_ids = tokenizer.texts_to_sequences(X_train)
    X_test_token_ids = tokenizer.texts_to_sequences(X_test)

    X_train_token_ids = tf.keras.preprocessing.sequence.
pad_sequences(X_train_token_ids, maxlen=sequence_length,
```

(continued)

Listing 9.7 (continued)
```
padding='post', value=0)
    X_test_token_ids = tf.keras.preprocessing.sequence.
pad_sequences(X_test_token_ids, maxlen=sequence_length,
padding='post', value=0)

    return X_train_token_ids, X_test_token_ids, tokenizer
```

Next, we create token IDs for the train and test set with a sequence length of 100:

```
sequence_length = 100
X_train_token_ids, X_test_token_ids, tokenizer = tokenization
(X_train, X_test, sequence_length)
```

9.5.2 Training

Now that we have tokenized the raw text into token IDs, we are ready to train the RNN model with embeddings. Let us first initialize the parameters. We will be learning 32-dimensional embedding vector for each word. The activation nodes are set to 10:

```
vocab_size = len(tokenizer.word_index)
n_x = vocab_size
n_y = 4
n_a = 10
n_e = 32

parameters = initialize_parameters(n_x, n_a, n_y, n_e)
```

Next, let us implement the end-to-end training for the RNN with the embedding layer:

Listing 9.8 Training Contextual Embedding Model
```
def train(X_token_ids, y, parameters, num_epochs=10):

    # Define the optimizer
    optimizer = tf.keras.optimizers.legacy.Adam()

    for i_epoch in range(num_epochs):
        mini_batches = create_mini_batches(X_token_ids, y,
batch_size=1024, seed=10)
```

(continued)

Listing 9.8 (continued)

```
        for mini_batch in mini_batches:
            X_token_ids_batch, y_batch = mini_batch

            with tf.GradientTape() as tape:
                # Watch parameters
                tape.watch(parameters)

                # Forward propagation
                X_embeddings = embedding_forward_propagation
(X_token_ids_batch, parameters['We'])
                A, Y_hat = rnn_forward_propagation(X_embeddings,
parameters)

                # Get y_hat for the non-padded last sequence
                mask = tf.math.not_equal(X_token_ids_batch, 0)
                seq_lengths = tf.math.reduce_sum(tf.cast(mask,
tf.int32), axis=-1)
                seq_lengths = seq_lengths - 1
                batch_indices = tf.range(start=0, limit=tf.shape
(Y_hat)[1], dtype=tf.int32)
                indices = tf.stack([batch_indices, seq_lengths],
axis=1)
                y_hat = tf.gather_nd(tf.transpose(Y_hat, perm=
[1, 2, 0]), indices)

                # Loss
                loss = tf.keras.losses.
sparse_categorical_crossentropy(y_batch, y_hat)

                loss = tf.reduce_mean(loss)

            # Compute gradients
            gradients = tape.gradient(loss, parameters.values())

            # Update weights
            optimizer.apply_gradients(zip(gradients, parameters.
values()))

        print('Epoch: %d, Loss: %f' % (i_epoch, loss.numpy()))

    return parameters
```

From the code provided, we can see that there are two distinct steps in the forward propagation process. The first step involves the embedding layer, while the second step pertains to the recurrent neural network (RNN). It is important to highlight that in this many-to-one RNN architecture, the predicted probability is derived from the last sequence that is not padded. The loss is then calculated based on this predicted probability, and subsequently, the gradients are computed in relation to this loss.

Let us now train for 30 epochs:

```
parameters = train(X_train_token_ids, y_train, parameters,
num_epochs=30)
```

9.5.3 *Prediction and Validation*

We know that model prediction is essentially the forward propagation steps. The following code implements the prediction step. Note that we are taking the last non-padded sequence as the final output:

Listing 9.9 Prediction Probabilities and Embeddings Computation

```
def predict(X_token_ids, parameters):

    # Forward propagation
    X_embeddings = embedding_forward_propagation(X_token_ids,
parameters['We'])
    A, Y_hat = rnn_forward_propagation(X_embeddings, parameters)

    # Get y_hat for the non-padded last sequence
    mask = tf.math.not_equal(X_token_ids, 0)
    seq_lengths = tf.math.reduce_sum(tf.cast(mask, tf.int32),
axis=-1)
    seq_lengths = seq_lengths - 1
    batch_indices = tf.range(start=0, limit=tf.shape(Y_hat)[1],
dtype=tf.int32)
    indices = tf.stack([batch_indices, seq_lengths], axis=1)
    y_hat = tf.gather_nd(tf.transpose(Y_hat, perm=[1, 2, 0]),
indices)

    return y_hat, tf.transpose(X_embeddings, [1, 2, 0])
```

Let us now execute the prediction code to compute prediction probabilities and embeddings:

```
y_hat, X_test_embeddings = predict(X_test_token_ids, parameters)
X_test_embeddings = X_test_embeddings.numpy()
y_pred = np.argmax(y_hat.numpy(), 1)
```

To evaluate the performance of the trained model on the test set, we can generate a classification report using the Scikit-learn library (Pedregosa et al. 2011) as follows:

```
from sklearn.metrics import classification_report
y_test = np.array(y_test)
print (classification_report(y_test, y_pred))
```

This prints:

```
              precision    recall  f1-score   support

           0       0.84      0.84      0.84      1900
           1       0.92      0.89      0.91      1900
           2       0.75      0.74      0.75      1900
           3       0.74      0.76      0.75      1900

    accuracy                           0.81      7600
   macro avg       0.81      0.81      0.81      7600
weighted avg       0.81      0.81      0.81      7600
```

This tells us that we have developed a successful classification system with embeddings.

9.5.4 Semantic Mapping

Let us now dive deeper into the learned embeddings. The t-SNE plot (Van der Maaten and Hinton 2008) serves as an excellent visualization tool, enabling us to better understand the embeddings and their correspondence to each news topic class. In the provided code below, we are using the t-SNE algorithm to reduce the dimensionality of our embeddings to 2D space. This transformation allows us to visualize the high-dimensional data in a 2D space while preserving the relationships between data points. Once we have the 2D embeddings, we create a scatter plot for each class label. For each unique label, we find the corresponding embeddings and plot them on the scatter plot. This gives us a visual representation of how the embeddings for each class are distributed in the 2D space:

Listing 9.10 Plotting t-SNE

```
import matplotlib.pyplot as plt
from sklearn.manifold import TSNE

def plot_tsne(embeddings, labels):

    # Use t-SNE to reduce dimensionality to 2D
    tsne = TSNE(n_components=2, perplexity=50, random_state=30)
    embeddings_2d = tsne.fit_transform(embeddings)

    # Create a scatter plot
    plt.figure()
    for i, label in enumerate(set(labels)):
        # Find points with this label
        indices = labels == label

        # Extract corresponding embeddings
        embeddings_iclass = embeddings_2d[indices]

        # Plot these embeddings
        plt.scatter(embeddings_iclass[:, 0], embeddings_iclass[:,
1], label=label, s=5)

    # Add a legend
    plt.legend()

    return None
```

Executing the plotting code will result in a t-SNE plot as shown in Fig. 9.3:

```
plot_tsne(X_test_embeddings_mean, y_test)
```

By examining this plot, we can gain insights into the semantic relationships between different news articles and how well our model has learned to differentiate between the various topic classes. We can see that news articles in the same class are grouped together in the embedding space. Each point in this high-dimensional embedding space represents a unique news article. The proximity of these points is indicative of their semantic similarity. In other words, articles that share similar themes or topics are positioned closer together, forming distinct clusters. This spatial grouping highlights the efficacy of our model in differentiating between various topic classes.

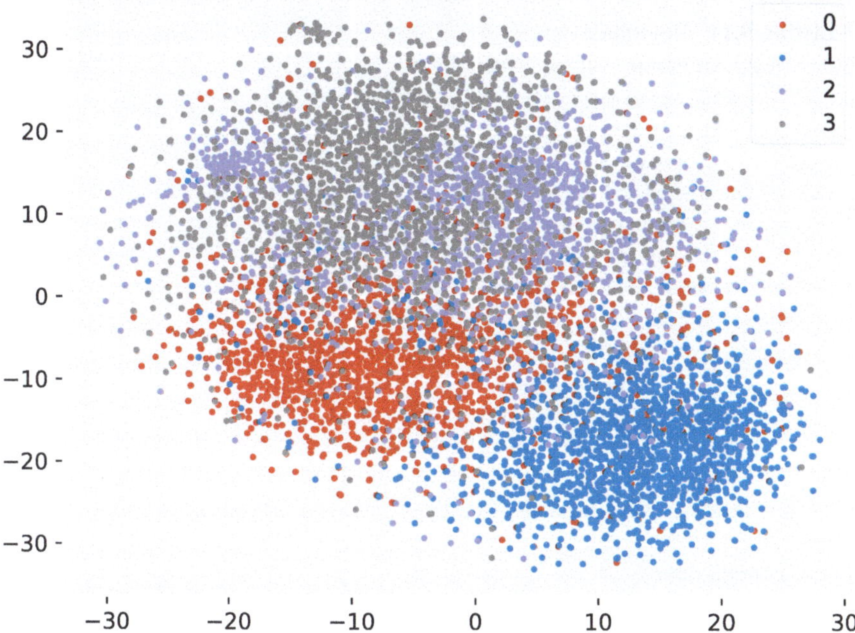

Fig. 9.3 The t-SNE plot of news topic categories

As previously mentioned, the learned embeddings from our model can serve a crucial role in recommending similar news articles to users. This is accomplished by calculating the cosine similarity between two vector embeddings. In the context of our model, each news article is represented as a point in the embedding space. The cosine similarity computes the cosine of the angle between two points, serving as a measure of how comparable the two articles are in terms of their content and context. Our objective is to identify a news article that resides closest in the embedding space to the article currently being read by the user. We do this by iterating over the embedding vectors of all available articles and calculating the cosine similarity with the current article's embedding vector. When the cosine similarity exceeds a certain threshold, it indicates that the two news articles share semantic similarity in the embedding space. Consequently, the article can be recommended to the user.

Following is an example code on how to implement this:

Listing 9.11 Calculating Embedding Similarity

```
def embedding_similarity(X_test_embeddings_mean, X_test, y_test):

    i = 0
    j = 1
```

(continued)

Listing 9.11 (continued)

```
    for i in range(100):
        cosine_sims = []
        for j in range(len(X_test)):
            cosine_sim = cosine_similarity(X_test_embeddings_mean
[i], X_test_embeddings_mean[j])
            cosine_sims.append(cosine_sim)
        idx = np.argsort(cosine_sims)[::-1][1]
        if cosine_sims[idx] < 0.8:
            continue
        print("--------------------------")
        print("Current News:", X_test[i])
        print("Current Label:", y_test[i])
        print("Similar News:", X_test[idx])
        print("Similar Label:", y_test[idx])
        print("Cosine Similarity:", cosine_sims[idx])

    return None
```

Here is an example output after running the above code:

```
Current News: Title: Sony NW-E95 and NW-E99 Network Walkman
Description: Sony Europe has launched two tiny 512MB and 1GB MP3 players,
the NW-E95 and NW-E99 Network Walkman. Both play MP3 (Sony has officially
bit the mp3 bullet) and ATRAC3plus compressed files and have a small blue
backlit LCD screen.
Current Label: 3
Similar News: Title: Sony launches music players with MP3 support
Description: The company has just announced the release of two flash-
memory-based devices, the Walkman NW-E99 and NW-E95, in Europe. The
music players can play songs in MP3 and Sony #39;s own Atrac file format.
Similar Label: 3
Cosine Similarity: 0.84968334
```

We can see that we have found a news article that is similar to the current article in the embedding space. Both articles discuss Sony and fall under the Sci/Tech category. The subject matter revolves around MP3 players. This similarity in content is reflected in the embedding space, where the two articles are positioned closely together. This demonstrates that our model is capable of identifying and recommending articles that are not just of similar topics but also contextually aligned.

9.6 Summary

- One-hot encoding for representing tokens or words treats each word as an isolated entity, failing to capture semantic and syntactic relationships.
- Word embeddings map words in a high-dimensional space, indicating semantic similarity through their locations and distances.
- Embedding representation generalizes better and captures nuanced information about words, improving predictions and enabling semantic analogies.
- An embedding layer converts tokenized input data into dense vectors known as embeddings using a lookup table and embedding matrix.
- The embedding matrix, a weight matrix in a neural network, is learned through backpropagation during training.
- Global embedding models generate single representations for each word, while contextual models, such as BERT and GPT, consider the surrounding context to produce different embeddings for the same word in different contexts.
- Cosine similarity measures semantic similarity between embedding vectors, with a range from -1 (opposite) to 1 (identical).
- Embeddings can be effectively learned through training on a task-specific problem, such as classification.

References

Bahdanau, D., Cho, K., & Bengio, Y. (2014). Neural machine translation by jointly learning to align and translate. *arXiv Preprint arXiv:1409.0473.*

Bojanowski, P., Grave, E., Joulin, A., & Mikolov, T. (2017). Enriching word vectors with subword information. *Transactions of the Association for Computational Linguistics*, *5*, 135–146.

Cer, D., Yang, Y., Kong, S., Hua, N., Limtiaco, N., John, R. S., Constant, N., Guajardo-Cespedes, M., Yuan, S., Tar, C., & others. (2018). Universal sentence encoder. *arXiv Preprint arXiv:1803.11175.*

Chollet, F. (2021). *Deep learning with Python*. simon and schuster.

Devlin, J., Chang, M.-W., Lee, K., & Toutanova, K. (2019). Bert: Pre-training of deep bidirectional transformers for language understanding. *Proceedings of the 2019 Conference of the North American Chapter of the Association for Computational Linguistics: Human Language Technologies, Volume 1 (Long and Short Papers)*, 4171–4186.

Goodfellow, I. (2016). *Deep learning*. MIT press.

Liu, Y., Ott, M., Goyal, N., Du, J., Joshi, M., Chen, D., Levy, O., Lewis, M., Zettlemoyer, L., & Stoyanov, V. (2019). Roberta: A robustly optimized BERT pretraining approach. *arXiv Preprint arXiv:1907.11692.*

Mikolov, T., Chen, K., Corrado, G., & Dean, J. (2013a). Efficient estimation of word representations in vector space. *arXiv Preprint arXiv:1301.3781.*

Mikolov, T., Sutskever, I., Chen, K., Corrado, G. S., & Dean, J. (2013b). Distributed representations of words and phrases and their compositionality. *Advances in Neural Information Processing Systems*, *26*.

Mikolov, T., Yih, W., & Zweig, G. (2013c). Linguistic regularities in continuous space word representations. *Proceedings of the 2013 Conference of the North American Chapter of the Association for Computational Linguistics: Human Language Technologies*, 746–751.

Pedregosa, F., Varoquaux, G., Gramfort, A., Michel, V., Thirion, B., Grisel, O., Blondel, M., Prettenhofer, P., Weiss, R., Dubourg, V., & others. (2011). Scikit-learn: Machine learning in Python. *The Journal of Machine Learning Research, 12*, 2825–2830.

Pennington, J., Socher, R., & Manning, C. D. (2014). Glove: Global vectors for word representation. *Proceedings of the 2014 Conference on Empirical Methods in Natural Language Processing (EMNLP)*, 1532–1543.

Peters, M. E., Neumann, M., Iyyer, M., Gardner, M., Clark, C., Lee, K., & Zettlemoyer, L. (2018). Deep Contextualized Word Representations. *ArXiv, abs/1802.05365*. https://api.semanticscholar.org/CorpusID:3626819

Radford, A., Narasimhan, K., Salimans, T., Sutskever, I., & others. (2018). *Improving language understanding by generative pre-training.*

Singhal, A. & others. (2001). Modern information retrieval: A brief overview. *IEEE Data Eng. Bull., 24*(4), 35–43.

Sutskever, I., Vinyals, O., & Le, Q. V. (2014). Sequence to sequence learning with neural networks. *Advances in Neural Information Processing Systems, 27.*

Van der Maaten, L., & Hinton, G. (2008). Visualizing data using t-SNE. *Journal of Machine Learning Research, 9*(11).

Vaswani, A., Shazeer, N., Parmar, N., Uszkoreit, J., Jones, L., Gomez, A. N., Kaiser, Ł., & Polosukhin, I. (2017). Attention is all you need. *Advances in Neural Information Processing Systems, 30.*

Zhang, X., Zhao, J., & LeCun, Y. (2015). Character-level convolutional networks for text classification. *Advances in Neural Information Processing Systems, 28.*

Chapter 10
Assembling Attention Mechanisms and Transformers

This chapter covers:

- Understanding Transformer model architecture.
- Exploring Byte Pair Encoding for efficient tokenization.
- Eliminating sequential dependencies through the self-attention mechanism.
- Diving into the core components: embeddings, encoder, decoder, and output layer.
- Illustrating a practical example: question and answering bot case study.
- Predicting tokens during inference with the Transformer.

ChatGPT and Gemini have become household names in the today's world. These are known as large language models (LLMs). The transformer architecture forms the backbone of these LLMs, driving their success. This chapter looks into the Transformer model's architecture, including its core components—embeddings, encoder, decoder, and output layer—illustrated through practical examples such as question and answering. We also explore essential techniques such as Byte Pair Encoding for efficient tokenization and the calculation of embeddings that capture both semantic meaning and positional information, setting the foundation for a deep understanding of the Transformer's inner workings.

10.1 Transformer Representation

In traditional sequence-to-sequence models, such as recurrent neural networks (RNNs) and long short-term memory (LSTM) networks (Hochreiter and Schmidhuber 1997; Lipton et al. 2015), the training process handles tokens in a sequential manner. This sequential dependency hinders the parallel computations. However, the Transformer model (Vaswani et al. 2017) does not have sequential dependencies. It processes all tokens in the input sequence in parallel during training. This parallel processing capability is primarily attributed to its self-attention

T. Islam, *Hands-on Deep Learning*, https://doi.org/10.1007/978-3-032-00488-8_10

mechanism. The self-attention mechanism enables the model to consider all tokens in the input sequence simultaneously, thereby eliminating the need for sequential dependencies.

It is important to note, however, that despite its parallel processing capabilities during training, the Transformer model still predicts tokens one at a time during the prediction phase. Despite this, the training characteristic of the Transformer model makes it exceptionally powerful, especially when paired with modern hardwares such as graphics processing units (GPUs) that are designed for high-speed parallel processing. This combination unlocks new levels of performance and efficiency, making the Transformer model a game-changer in the field of sequence-to-sequence modeling.

The Transformer model architecture consists of following four core components (Fig. 10.1):

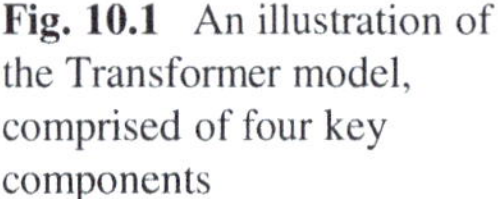

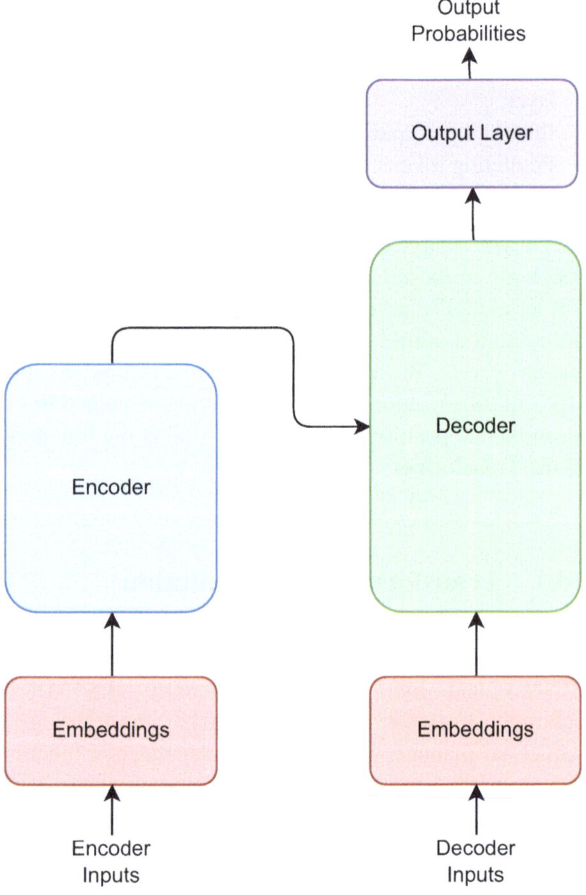

Fig. 10.1 An illustration of the Transformer model, comprised of four key components

- Embeddings
- Encoder
- Decoder
- Output layer

Let us take a look at an example regarding English to French translation. We have the following sentence in English:

Taylor Swift performs in Portugal for first time

We also have the label of French translation:

Taylor Swift se produit pour la première fois au Portugal

In the context of Transformer models, the training process involves both an encoder and a decoder. The encoder processes the input data (e.g., the English sentence), and the decoder generates the output data (e.g., French translation). If we want to train and apply a Transformer model, the encoder input in this case will be the English sentence, that is *Taylor Swift performs in Portugal for first time*. The decoder input will be shifted right of the French translation label. The "right shifting" refers to the process of shifting the decoder input one step to the right. Therefore, in this case:

- The decoder input (right-shifted) would be: "<start> Taylor Swift se produit pour la première fois au Portugal"
- The decoder target (output) would be: "Taylor Swift se produit pour la première fois au Portugal <end>"

Here, "<start>" and "<end>" are special tokens indicating the beginning and end of a sentence, respectively. So, the model is always trying to predict the next word in the sentence given the previous words. It does not have access to the future words during training, which is why the input is "shifted right." This technique is also known as "teacher forcing" (Williams and Zipser 1989).

Figure 10.2 provides detailed transformer architecture as seen in the original paper. We will be covering all these concepts in detail.

The embeddings component is comprised of two main elements:

- *Token embeddings*: These are used to represent individual tokens or words in a high-dimensional space.
- *Position embeddings*: These capture the position of each token within the sequence, providing contextual information of positions of the tokens to the model.

The encoder and decoder components consist of several key parts:

- *Attention mask*: This is used to prevent attention to certain tokens during the self-attention process.
- *Multi-head attention*: This mechanism allows the Transformer model to direct its attention to various positions simultaneously, thereby capturing a diverse range of information aspects.

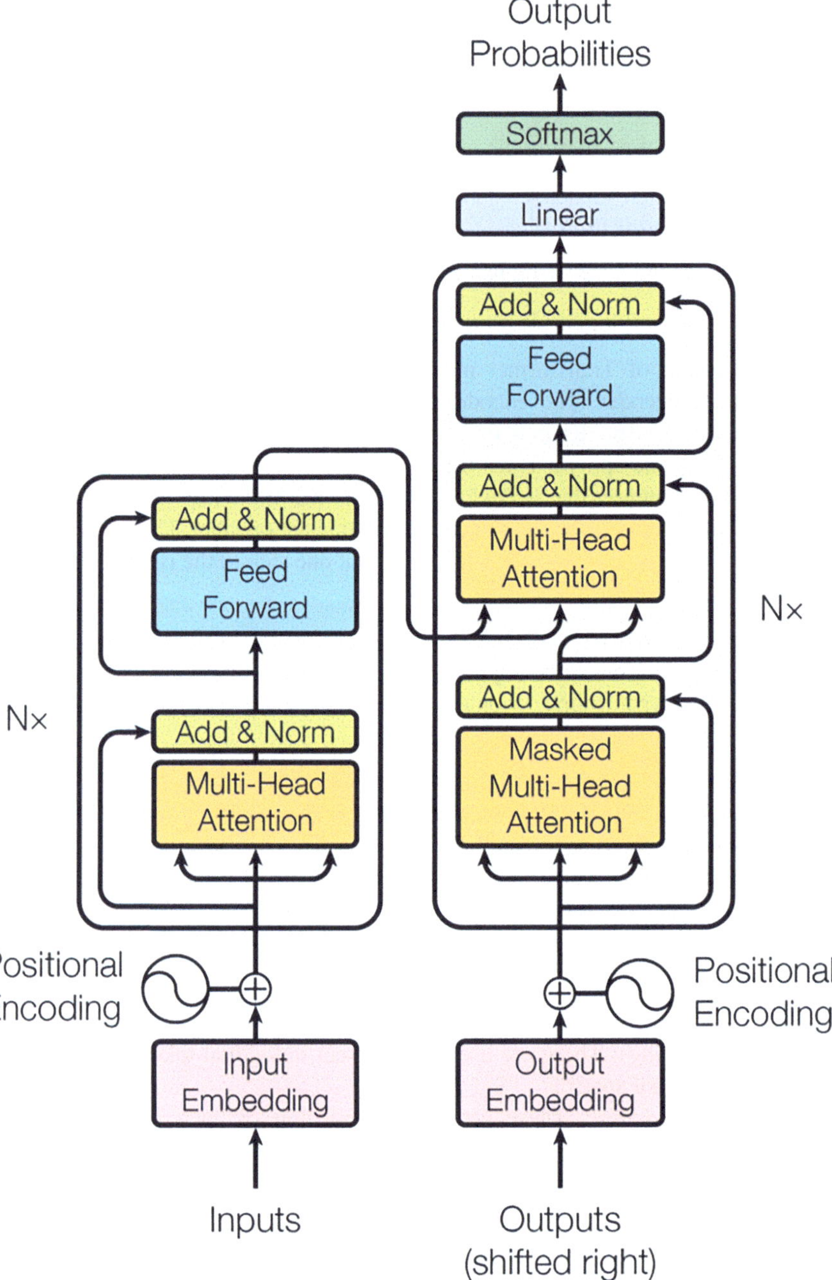

Fig. 10.2 An illustration of the Transformer architecture

- *Feed forward network*: This is a simple fully connected neural network that is applied to each position separately and identically.
- *Add & Norm*: This includes residual connections (Add) followed by layer normalization (Norm), which helps in stabilizing the learning process and allows for deeper models.

10.2 Byte Pair Encoding

10.2.1 Overview

Before discussing the Transformer models, let us first discuss a tokenization method called Byte Pair Encoding (BPE) (Sennrich et al. 2015), as it is often used in conjunction with Transformer models. Referring back to the transformer representation as shown in Figs. 10.1 and 10.2, we start with inputs. We know that the initial step of any natural language task is tokenization, which involves breaking down the text into individual tokens. Previously, we looked at character-level and word-level tokenization. However, character-level and word-level tokenization have their limitations.

Word-level tokenization can lead to a very large vocabulary that includes all the unique words in the corpus. Imagine, if we are training modern LLM (Brown et al. 2020) on internet data, there will be millions of tokens using word-level tokenization. This will be computationally expensive. Word-level tokenization also suffers from the issue of out-of-vocabulary (OOV) words. If a word is not seen in the training data, the model will not be able to handle it during inference. Character-level tokenization solves this problem to some extent, as it generates a small vocabulary size. However, it also results in a longer sequence length, which presents its own set of challenges.

BPE is a good middle ground as it combines the advantages of both word-level and character-level tokenization. It is a type of subword tokenization method. The idea behind BPE is to iteratively replace the most frequent pair of characters or character sequences in a text with a new token ID. This process effectively reduces the size of the vocabulary, allowing the model to handle a wider range of texts and tokens.

Imagine we have a corpus with words such as "running," "walking," "crying," "feeding," and so on. We notice that the letters "ing" appears very often. To save space, we could replace every "ing" with a new token that is not used anywhere else in the corpus. We will also add this token in the vocabulary to keep track that says this new token actually represents "ing." This way, we can compress the corpus without losing any information! That is essentially what Byte Pair Encoding does.

10.2.2 Learning

Here is how the Byte Pair Encoding (BPE) learns from training corpus in a series of steps in iterative fashion:

Initialization

We start with a base vocabulary of individual characters present in the training corpus. The initial vocabulary in BPE is constructed from the set of all unique characters present in the corpus. This is how we can codify the initialization:

Listing 10.1 Initialization of Vocabulary in BPE

```
def initialize_vocabulary(text):
    # Create a set of unique characters
    unique_chars = sorted(set(text))

    unknown_token = '<UNK>'
    unique_chars.append(unknown_token)

    # Create a dictionary where each unique character is associated
with a unique ID
    vocab = {char: id for id, char in enumerate(unique_chars)}
    reverse_vocab = {id: char for char, id in vocab.items()}

    return vocab, reverse_vocab
```

Pair Statistics

Next, the frequency of each character pair in the corpus is calculated. For example, the word “loneliness” will generate the following pair count:

```
{('l', 'o'): 1,
('o', 'n'): 1,
('n', 'e'): 2,
('e', 'l'): 1,
('l', 'i'): 1,
('i', 'n'): 1,
('e', 's'): 1,
('s', 's'): 1}
```

In practice, we compute the pair statistics after converting characters into token IDs, This is how we can do this:

Listing 10.2 Pair Statistics Computation in BPE

```
def get_stats(token_ids, pair_count=None):
    if pair_count is None:
        pair_count = {}
    for id_pair in zip(token_ids, token_ids[1:]):
        if id_pair not in pair_count:
            pair_count[id_pair] = 1
        else:
            pair_count[id_pair] += 1
    return pair_count
```

Merge Operation

Next, we find the most frequent pair of tokens and merge them to form a new token. This new token will be added to the vocabulary. The merge operation will look like this to produce new list of token IDs:

Listing 10.3 Merge Operation in BPE

```
def merge(token_ids, pair, new_idx):
    new_token_ids = []
    skip_next = False

    for i in range(len(token_ids)):
        if skip_next:
            skip_next = False
            continue

        if i < len(token_ids) - 1 and (token_ids[i], token_ids[i + 1])
== pair:
            new_token_ids.append(new_idx)
            skip_next = True
        else:
            new_token_ids.append(token_ids[i])

    return new_token_ids
```

Repeat Pair Statistics and Merge Operations

We repeat the pair statistics and merge operation steps until we reach a predefined number of merges or until no more merges can be performed. The iteration loop will look like this:

```
for i in range(num_merges):
    pair_count = get_stats(token_ids)
    best_pair = max(pair_count, key=pair_count.get)

    current_vocab_size = max(reverse_vocab)
    new_vocab_id = current_vocab_size + 1
    token_ids = merge(token_ids, best_pair, new_vocab_id)
```

It is important to note that the number of merges is a hyperparameter that we can tune. A larger number of merges will result in a larger vocabulary with longer tokens, while a smaller number of merges will result in a smaller vocabulary with shorter tokens. The optimal number of merges depends on the specific task and dataset.

10.2.3 Fitting, Encoding, and Decoding

Now that we have covered the fundamentals of learning BPE, let us put all the pieces together to create the BPE tokenizer. This is shown in the code as follows:

Listing 10.4 Byte-Pair Encoder

```
class BytePairEncoder:
    def __init__(self):
        super().__init__()

    def fit(self, text, num_merges):

        vocab, reverse_vocab = initialize_vocabulary(text)

        token_ids = list(map(vocab.get, text))

        merges = {}
        for i in range(num_merges):
            pair_count = get_stats(token_ids)
            best_pair = max(pair_count, key=pair_count.get)
```

(continued)

Listing 10.4 (continued)

```
            current_vocab_size = max(reverse_vocab)
            new_vocab_id = current_vocab_size + 1
            token_ids = merge(token_ids, best_pair, new_vocab_id)

            merges[best_pair] = new_vocab_id
            reverse_vocab[new_vocab_id] = reverse_vocab[best_pair
[0]] + reverse_vocab[best_pair[1]]
            vocab = {v: k for k, v in reverse_vocab.items()}

        self.merges = merges
        self.vocab = vocab
        self.reverse_vocab = reverse_vocab

    def encode(self, text):
        token_ids = [self.vocab.get(token, self.vocab.get
('<UNK>')) for token in text]
        new_token_ids = token_ids
        for pair, new_idx in self.merges.items():
            new_token_ids = merge(new_token_ids, pair, new_idx)
        return new_token_ids

    def decode(self, token_ids):
        decoded_text = ''.join(self.reverse_vocab[id] for id in
token_ids)
        return decoded_text
```

The above `BytePairEncoder` class is an implementation of the BPE algorithm. It has methods for training the encoder (`fit`), encoding text (`encode`), and decoding token IDs (`decode`). The `fit` method is where the BPE model is trained. It starts with a basic vocabulary initialized from the input text. Then it performs a number of merge operations, each time finding the most frequent pair of tokens in the text and merging them into a new token. This new token is added to the vocabulary, and the text is updated to reflect this merge. The `encode` method takes a piece of text and encodes it into a sequence of token IDs. It does this by replacing each token in the text with its corresponding ID from the vocabulary. If a token is not in the vocabulary, it gets replaced with the ID for an unknown token. The method then enters a loop where it iteratively applies the merges that were learned during the training phase. For each pair of tokens in the `merges` dictionary, the method looks for occurrences of that pair in the text and replaces them with the ID of the merged token. This process continues until no more replaceable pairs are left or the pair is not in the vocabulary. The `decode` method does the opposite of encode. It takes a sequence of token IDs and replaces each ID with its corresponding token from the vocabulary, effectively reconstructing the original text.

Here is an example of applying the fit, encode, and decode methods:

```
bpe = BytePairEncoder()
text = "Let's assume this is our corpus."

bpe.fit(text, 3)
token_ids = bpe.encode("Let's check!")
decoded_text = bpe.decode(token_ids)
print (token_ids)
print (decoded_text)
```

This prints:

```
[19, 14, 1, 17, 5, 7, 6, 5, 16, 16]
Let's chec<UNK><UNK>
```

10.3 Embeddings

Following tokenization, we compute the embeddings for each token in the sequence. These embeddings are calculated based on a learned embedding weight matrix.

When it comes to the Transformer model, there are two types of embeddings that we need to compute and learn. The first type is the "token embeddings," which represent the semantic meaning of each individual token (Press and Wolf 2016). The second type is the "position embeddings," which capture the positional information of each token within the sequence (Gehring et al. 2017). This dual-embedding approach is a key feature of the Transformer model, enabling it to effectively understand and generate natural language. The visual illustration of the embeddings representation is shown in Fig. 10.3.

Token embeddings are computed using a lookup table that stores the token embedding weights matrix. We have already covered embeddings, which is essentially token embeddings, in detail in the previous chapter.

We will now take a look at an example. In this example, we are assuming the following attributes:

```
batch_size = 32
sequence_length = 50
embedding_dim = 256
vocab_size = 30000
```

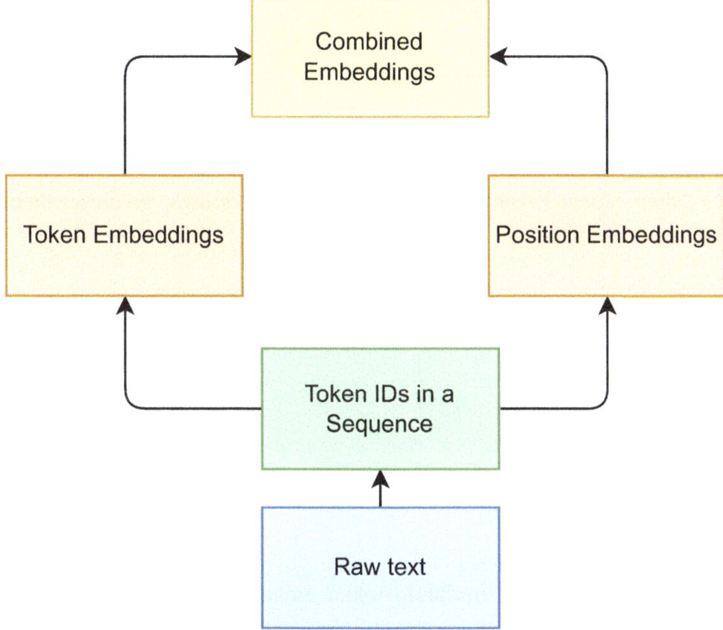

Fig. 10.3 A visual illustration of the embeddings representation in the Transformer model

First, we will generate a matrix composed of random sequences. To accommodate sequences of varying lengths, we will apply padding to some of them. We can implement this as follows:

Listing 10.5 Creating Random Sequences

```
def create_random_sequences(batch_size, sequence_length):

    X_sequences = tf.random.uniform([batch_size,
sequence_length], minval=0, maxval=vocab_size, dtype=tf.
int32).numpy()

    for i in range(batch_size):
        pad_size = np.random.randint(0, sequence_length//2)
        X_sequences[i, -pad_size:] = 0

    return X_sequences
```

We can execute the above function as follows:

```
batch_size = 32
sequence_length = 50
X_sequences = create_random_sequences(batch_size, sequence_length)
```

We can then proceed to calculate the token embeddings, given an embedding weight matrix, as follows:

```
W_token_embeddings = tf.random.uniform([vocab_size, embedding_dim]).numpy()
X_token_embeddings = W_token_embeddings[X_sequences]
```

Similarly, we can also apply TensorFlow's `embedding_lookup` to achieve this:

```
X_token_embeddings = tf.nn.embedding_lookup(W_token_embeddings,
X_sequences)
```

Position embeddings are similar to token embeddings. However, while token embeddings learn representations for each token in the vocabulary, position embeddings learn representations for different positions in a sequence. Let us denote the position embeddings as P. If we have a sequence of length n, then P is an $n \times d$ matrix, where d is the dimension of the embeddings. Each row of P corresponds to the position embedding of a particular position in the sequence. The equation for position embedding computation can be written as:

$$P(j) = W_P[j]$$

Here is the code to compute position embeddings given a position embedding weight matrix:

```
# Positional embeddings weights
W_position_embeddings = tf.Variable(tf.random.uniform
([sequence_length, embedding_dim]))

# Create a matrix of positional indices
positions = tf.range(sequence_length)
positions = tf.expand_dims(positions, 0)
positions = tf.repeat(positions, batch_size, axis=0)

# Get the positional embeddings
X_position_embeddings = tf.nn.embedding_lookup
(W_position_embeddings, positions)
print (X_position_embeddings.shape)
(32, 50, 256)
```

The final embeddings are essentially the result of adding the token embeddings and the position embeddings together:

```
X_embeddings = X_token_embeddings + X_position_embeddings
```

10.4 Encoder and Decoder Components

10.4.1 Attention Mask

The attention mask essentially is a binary matrix that indicates which tokens the model should be attended to and which tokens should be ignored. In the matrix, the value 1 represents positions that should be attended to, and 0 denotes positions to be ignored or masked. By applying an attention mask, the model can focus on the meaningful tokens and disregard the other tokens. There are two types of attention masks used in a transformer: padding mask and look-ahead mask.

Padding mask: The padding mask is used to ensure that the model does not attend to padded tokens as they do not contain any meaningful information. The following code example shows how to create a padding mask:

Listing 10.6 Creating Padding Mask

```
def create_padding_mask(X_sequences):

    X_padding_mask = tf.cast(tf.math.equal(X_sequences, 0), tf.
float32)
   X_padding_mask = X_padding_mask[:, tf.newaxis, tf.newaxis, :]

    return X_padding_mask
```

Look-ahead mask: The look-ahead mask is used to mask future tokens in a sequence. This means that to predict the second token, only the first token is used. Similarly to predict the third token, only the first and second tokens will be used and so on. This ensures that the model is not able to "cheat" by looking at future tokens in the target sequence during training. The implementation of look-ahead mask is provided as follows:

Listing 10.7 Creating Look Ahead Mask

```
def create_look_ahead_mask(sequence_length):

    X_look_ahead_mask = 1 - tf.linalg.band_part(tf.ones
((sequence_length, sequence_length)), -1, 0)
    X_look_ahead_mask = X_look_ahead_mask[tf.newaxis, tf.
newaxis, :, :] # (batch_size, 1, seq_len, seq_len)

    return X_look_ahead_mask
```

Finally, we combine the padding mask and look-ahead mask in the following code. Note that, in encoder, we only use padding mask as the attention mask. In decoder's first sub-layer, the look-ahead mask is also used. In the second sub-layer of the decoder, only the padding mask is used. The look-ahead mask is not used in the second sub-layer of the decoder. The look-ahead mask is used in the decoder's first sub-layer to prevent the model from seeing future tokens in its own sequence, but it's not needed in the second sub-layer because that layer attends to encoder outputs, which have no "future" constraint. This can be configured via `use_look_ahead` flag:

Listing 10.8 Creating Attention Mask by Combining Padding and Look Ahead Masks

```
def create_attention_mask(X_sequences, use_look_ahead=False):

    sequence_length = tf.shape(X_sequences)[1]
    padding_mask = create_padding_mask(X_sequences)

    if use_look_ahead:
        look_ahead_mask = create_look_ahead_mask
(sequence_length)
        X_attention_mask = tf.maximum(look_ahead_mask,
padding_mask) # (batch_size, 1, seq_len, seq_len)
    else:
        X_attention_mask = padding_mask

    return X_attention_mask
```

We can run the preceding code as follows:

```
X_attention_mask = create_attention_mask(X_sequences,
use_look_ahead=True)
print (X_attention_mask.shape)
```

```
(32, 50, 256)
```

10.4.2 Multi-head Attention

Figure 10.4 provides a step-by-step flow of the multi-head attention layer. It starts with embeddings. Then, we compute the Q (Query), K (Key), and V (Value) matrices. These matrices are computed by multiplying the input embeddings with their respective weight matrices:

$$Q = X \cdot W_q$$

$$K = X \cdot W_k$$

$$V = X \cdot W_v$$

where:

- X is the combined token and position embeddings
- W_q, W_k, and W_v are the weight matrices for Query, Key, and Value, respectively.

Next, we partition each of the Q (Query), K (Key), and V (Value) matrices into multiple "heads." This implies that the embedding dimension of these matrices is divided into a number of segments equal to the predefined hyperparameter known as the "number of heads." To illustrate, if we decide to split these matrices into 8 heads, a Q matrix with dimensions for batch size, sequence length, and embedding dimension represented as (32, 50, 256) would be reshaped into a 4D tensor with dimensions (32, 50, 8, 32). This transformation would allow the model to focus on different parts of the input sequence simultaneously, capturing various types of information and patterns.

Conceptually, the multi-head attention mechanism in Transformer models can be related to the operation of convolutional neural networks (CNNs). In CNNs, different filters learn to recognize various features in the data, such as edges and textures. Similarly, in the context of multi-head attention, each "head" can be thought of as a unique "filter" that focuses on different aspects of the input sequence. For instance, one head might concentrate on identifying "who" is involved in a given context, while another might specialize in discerning "what" is the context, and so forth. This allows the model to capture a diverse range of information from the input, enhancing its ability to understand and generate meaningful outputs.

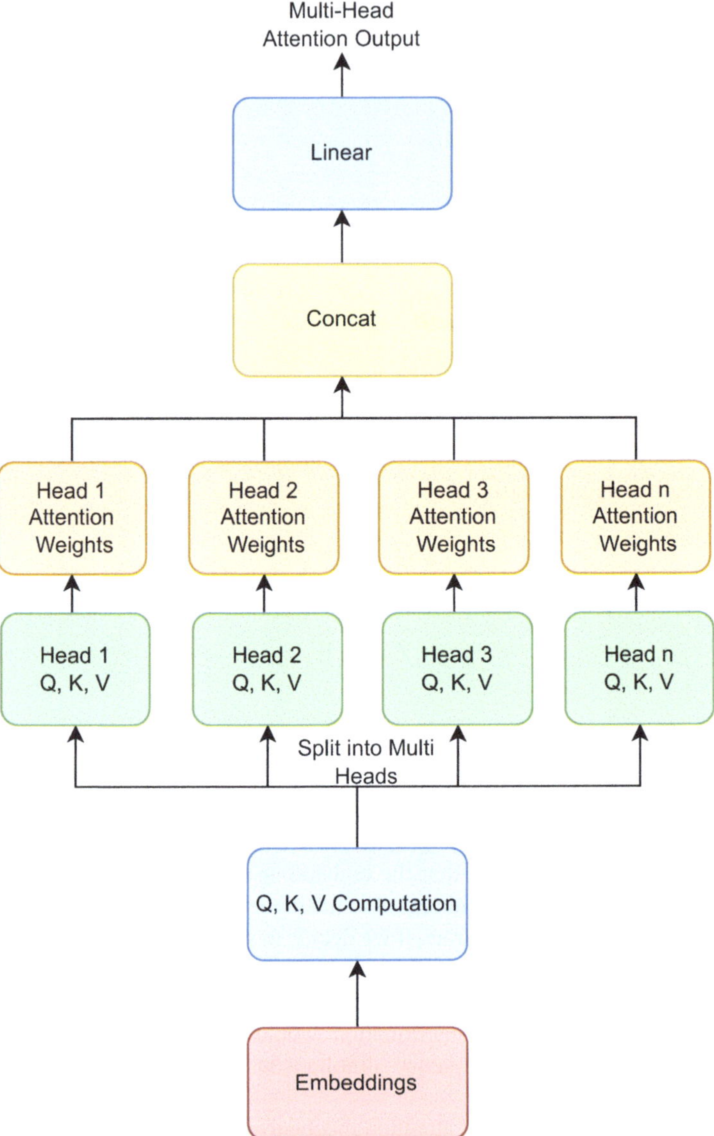

Fig. 10.4 Multi-head attention mechanism in the Transformer model

Next, the "scaled dot-product attention" is calculated for each head. This involves taking the dot product of the Q and K matrices, scaling it by the square root of the dimension of the key vectors, incorporating mask if needed, and then applying a `softmax` function to obtain the attention weights. These attention weights are then multiplied with the V matrix to get the output of each head:

$$\text{Attention}\ (Q, K, V) = softmax\left(\frac{QK^T}{\sqrt{d_k}} + M\right)V$$

In the scaled dot-product attention, d_k is the dimension of the key vectors within a single attention head, where $d_k = d_{model}$/num_heads = embedding_dimension/num_heads. It is also the dimension of the query vectors, as they must have the same dimension for their dot product to be meaningful. Dividing by $\sqrt{d_k}$ normalizes the variance of the dot products, stabilizing training by preventing values from becoming too large before the softmax. The mask matrix M is added to the scaled dot product (rather than multiplied) because its purpose is to effectively zero out attention weights. By placing very large negative values in M at positions to be masked, and adding M to the logits causes $e^{-\infty}$ to become 0 after the softmax function, thus ensuring those attention weights are effectively zero.

The outputs from all the heads are then concatenated and linearly transformed through a dense layer to produce the final output of the multi-head attention layer.

Here is the code that implements computation of multi-head attention output:

Listing 10.9 Multi-head Attention Layer

```
def multi_head_attention(X_embeddings, X_attention_mask,
num_heads, embedding_dim, parameters):

    # Depth of each attention head
    depth = embedding_dim // num_heads

    # Compute Q, K, V
    Q = tf.matmul(X_embeddings, parameters['W_q'])
    K = tf.matmul(X_embeddings, parameters['W_k'])
    V = tf.matmul(X_embeddings, parameters['W_v'])

    # Split Q, K, V into multiple heads
    Q = tf.reshape(Q, [batch_size, sequence_length, num_heads,
depth])
    K = tf.reshape(K, [batch_size, sequence_length, num_heads,
depth])
    V = tf.reshape(V, [batch_size, sequence_length, num_heads,
depth])

    # Transpose for computation of dot products and applying masks
    Q = tf.transpose(Q, perm=[0, 2, 1, 3])
    K = tf.transpose(K, perm=[0, 2, 1, 3])
    V = tf.transpose(V, perm=[0, 2, 1, 3])
```

(continued)

Listing 10.9 (continued)

```
    # Scaled dot-product attention
    matmul_qk = tf.matmul(Q, K, transpose_b=True)
    dk = tf.cast(tf.shape(K)[-1], tf.float32)
    attention_scores = matmul_qk / tf.math.sqrt(float(dk))

    # Add mask to the scaled tensor
    if X_attention_mask is not None:
        attention_scores += (X_attention_mask * -1e9)

    # Apply softmax to get attention weights
    attention_weights = tf.nn.softmax(attention_scores,
axis=-1)

    # Compute the output
    mh_attention_output = tf.matmul(attention_weights, V)
    mh_attention_output = tf.transpose(mh_attention_output,
perm=[0, 2, 1, 3])

    # Concatenate the multi-head outputs
    mh_attention_output = tf.reshape(mh_attention_output,
[batch_size, sequence_length, embedding_dim])

    # Output linear layer
    mh_attention_output = tf.matmul(mh_attention_output,
parameters['W_o'])

    return mh_attention_output
```

The above function can be executed by providing the weight matrices as follows:

```
# Define the weights for the Query, Key, Value, and Output matrices
parameters = {}
parameters['W_q'] = tf.Variable(tf.random.uniform([embedding_dim,
embedding_dim]))
parameters['W_k'] = tf.Variable(tf.random.uniform([embedding_dim,
embedding_dim]))
parameters['W_v'] = tf.Variable(tf.random.uniform([embedding_dim,
embedding_dim]))
parameters['W_o'] = tf.Variable(tf.random.uniform([embedding_dim,
embedding_dim]))
```

```
num_heads = 8 # Number of attention heads
mh_attention_output = multi_head_attention(X_embeddings,
X_attention_mask, num_heads, embedding_dim, parameters)
print (mh_attention_output.shape)
(32, 50, 256)
```

Referring to Fig. 10.4, once the multi-head attention output has been computed, the next step is to apply the Add & Normalize operation. This operation involves adding the inputs to the multi-head attention output. This is also known as residual connection. This is the "Add" part of the operation. Following this addition, normalization is performed to ensure that the output values are scaled appropriately. This is the "Normalize" part of the operation. This is to speed up the training. The implementation of this operation to multi-head attention will look like this:

Listing 10.10 Add & Norm Operation to Multi-head Attention

```
def mh_attention_add_norm(mh_attention_output, X_embeddings):
    mh_attention_add_output = mh_attention_output + X_embeddings
# Add
     mh_attention_add_norm_output = tf.keras.layers.
LayerNormalization()(mh_attention_output) # Norm

     return mh_attention_add_norm_output
```

We can run the above code as follows:

```
mh_attention_add_norm_output = mh_attention_add_norm
(mh_attention_output, X_embeddings)
print (mh_attention_output.shape)
(32, 50, 256)
```

10.4.3 Feedforward Network

Next, the output from the multi-head attention mechanism, after undergoing an Add & Normalize operation, is passed through a fully connected feed forward network:

Listing 10.11 Feedforward Layer

```
def feed_forward(mh_attention_add_norm_output, parameters):

    feed_forward_output = tf.nn.relu(tf.matmul
(mh_attention_add_norm_output, parameters['W_ff1']))
    feed_forward_output = tf.matmul(feed_forward_output,
parameters['W_ff2'])

    return feed_forward_output
```

We can compute the feed-forward output providing feed-forward weight matrices as follows:

```
parameters['W_ff1'] = tf.Variable(tf.random.uniform
([embedding_dim, feedforward_dim]))
parameters['W_ff2'] = tf.Variable(tf.random.uniform
([feedforward_dim, embedding_dim]))
feed_forward_output = feed_forward(mh_attention_add_norm_output,
parameters)
print (feed_forward_output.shape)
(32, 50, 256)
```

Similar to multi-head attention output, Add & Norm is also applied to feed-forward output:

Listing 10.12 Add & Norm Operation to Feed-Forward Network

```
def feed_forward_add_norm(mh_attention_add_norm_output,
feed_forward_output):
    feed_forward_add_output = mh_attention_add_norm_output +
feed_forward_output
    feed_forward_add_norm_output = tf.keras.layers.
LayerNormalization()(feed_forward_add_output)

    return feed_forward_add_norm_output
```

```
feed_forward_add_norm_output = feed_forward_add_norm
(mh_attention_add_norm_output, feed_forward_output)
print (feed_forward_add_norm_output.shape)
(32, 50, 256)
```

10.5 Encoder

Now that we have all the elements of an encoder, we can now proceed to construct the encoder. Here is the code of an encoder:

Listing 10.13 Encoder in Transformer

```
@tf.keras.saving.register_keras_serializable()
class TransformerEncoder(tf.keras.layers.Layer):
    def __init__(self, embedding_dim, num_heads, feedforward_dim,
rate=0.1):
        super(TransformerEncoder, self).__init__()
        self.multi_head_attention = tf.keras.layers.
MultiHeadAttention(num_heads=num_heads,
key_dim=embedding_dim)

        self.feed_forward = tf.keras.Sequential(
        [tf.keras.layers.Dense(feedforward_dim,
activation="relu"), tf.keras.layers.Dense(embedding_dim),]
        )
        self.layernorm1 = tf.keras.layers.LayerNormalization
(epsilon=1e-6)
        self.layernorm2 = tf.keras.layers.LayerNormalization
(epsilon=1e-6)
        self.dropout1 = tf.keras.layers.Dropout(rate)
        self.dropout2 = tf.keras.layers.Dropout(rate)

    def call(self, inputs):
        mh_attention_output = self.multi_head_attention(inputs,
inputs, inputs, use_causal_mask=False)
        mh_attention_output = self.dropout1(mh_attention_output)
        mh_attention_add_norm_output = self.layernorm1(inputs +
mh_attention_output)
        feed_forward_output = self.feed_forward
(mh_attention_add_norm_output)
        feed_forward_output = self.dropout2(feed_forward_output)
        encoder_output = self.layernorm2
(mh_attention_add_norm_output + feed_forward_output)
        return encoder_output
```

The input to the encoder is a combination of token and position embeddings as we discussed earlier. These embeddings encapsulate the semantic and syntactic information of the input language, providing a rich, high-dimensional representation of the source sentence.

There are two sub-layers in an encoder. The first sub-layer is a multi-head attention layer. The combined input embeddings is processed through a multi-head attention mechanism. This enables the model to simultaneously concentrate on different segments of the input sequence, thereby capturing diverse facets of the sequence's structure and significance. It is important to note that, in the encoder, we do not use a look-ahead mask. Instead, we only apply a padding mask to ignore irrelevant tokens, such as padding tokens added to equalize the lengths of sentences in a batch. This is because the encoder only receives inputs and does not need to predict future tokens, eliminating the possibility of "cheating" or prematurely accessing future information. For instance, consider a language translation task from English to French. The encoder inputs in this case would be the entire English sentence. There is no need for a look-ahead mask because the encoder is not required to generate predictions based on this input. Following the multi-head attention mechanism, the output is passed through an Add & Norm (Addition and Normalization) operation. This layer normalization (Ba et al. 2016) helps to stabilize the outputs of the multi-head attention and prepares them for the next stage of processing.

The second sub-layer is a feed-forward layer. The output from the Add & Norm from the first sublayer is fed into a feed-forward network in this sub-layer. This network further processes the information, extracting complex patterns and relationships. The output of this feed-forward network is again normalized using an Add & Norm operation. This is the final output of the encoder.

10.6 Decoder

The following is an implementation of a decoder for transformer:

Listing 10.14 Decoder in Transformer

```
@tf.keras.saving.register_keras_serializable()
class TransformerDecoder(tf.keras.layers.Layer):
   def __init__(self, embedding_dim, num_heads, feedforward_dim,
rate=0.1):
        super(TransformerDecoder, self).__init__()

        self.multi_head_attention1 = tf.keras.layers.
MultiHeadAttention(num_heads=num_heads,
key_dim=embedding_dim)
        self.multi_head_attention2 = tf.keras.layers.
MultiHeadAttention(num_heads=num_heads,
key_dim=embedding_dim)
```

(continued)

Listing 10.14 (continued)

```
        self.feed_forward = tf.keras.Sequential(
            [tf.keras.layers.Dense(feedforward_dim,
activation="relu"), tf.keras.layers.Dense(embedding_dim),]
        )

        self.layernorm1 = tf.keras.layers.LayerNormalization
(epsilon=1e-6)
        self.layernorm2 = tf.keras.layers.LayerNormalization
(epsilon=1e-6)
        self.layernorm3 = tf.keras.layers.LayerNormalization
(epsilon=1e-6)

        self.dropout1 = tf.keras.layers.Dropout(rate)
        self.dropout2 = tf.keras.layers.Dropout(rate)
        self.dropout3 = tf.keras.layers.Dropout(rate)

    def call(self, inputs, encoder_outputs):
        look_ahead_mask = create_attention_mask(inputs,
use_look_ahead=False)
        mh_attention_output1 = self.multi_head_attention1
(inputs, inputs, inputs, use_causal_mask=True)
        mh_attention_output1 = self.dropout1
(mh_attention_output1)
        mh_attention_add_norm_output1 = self.layernorm1(inputs +
mh_attention_output1)

        mh_attention_output2 = self.multi_head_attention2
(mh_attention_add_norm_output1, encoder_outputs,
encoder_outputs, use_causal_mask=False)
        mh_attention_output2 = self.dropout2
(mh_attention_output2)
        mh_attention_add_norm_output2 = self.layernorm2
(mh_attention_add_norm_output1 + mh_attention_output2)

        feed_forward_output = self.feed_forward
(mh_attention_add_norm_output2)
        feed_forward_output = self.dropout3(feed_forward_output)
        decoder_output = self.layernorm3
(mh_attention_add_norm_output2 + feed_forward_output)

        return decoder_output
```

The decoder mirrors the steps of the encoder, but incorporates an additional sub-layer, making a total of three. The process begins with the decoder input embeddings, which are a right-shifted version of the target output embeddings. These embeddings are then fed into the first sub-layer. In this initial sub-layer, a look-ahead mask is applied to prevent the model from attending to future positions. The second sub-layer also employs multi-head attention, but it computes this attention over the output of the encoder stack, creating a bridge between the encoder and decoder components. The third and final sub-layer is a feed-forward layer, which processes the information from the previous layers. Each of these three sub-layers is subjected to an Add & Norm operation, which helps to stabilize the outputs and improve the model's learning dynamics. The output from this third sub-layer is then considered as the final decoder output, marking the end of the decoding process.

10.7 Transformer Model

We will now create a Transformer model, stitching the encoder–decoder components that we just discussed. Before that, let us first walk through the dimensions of a Transformer model for the reference (Table 10.1).

Keeping the dimensions provided in the table in mind, let us now create the Transformer model. First, we will have to define the loss function. In language modeling, masked sparse categorical cross-entropy loss is a common choice. This loss function is specifically designed for sequence prediction tasks where the sequences can be of different lengths and we may want to mask certain parts of the output.

The "masked" part of the loss function refers to the ability to ignore certain parts of the output sequence during loss calculation. For example, in sequence prediction tasks, we often pad shorter sequences with a special padding token so that all sequences in a batch have the same length. However, we do not want these padding tokens to contribute to the loss, so we "mask" them out. We can codify this as follows:

Listing 10.15 Masked Sparse Categorical Cross-entropy Loss

```
@tf.keras.saving.register_keras_serializable()
def masked_sparse_categorical_crossentropy(labels,
predictions):

    # Create a mask for non-zero (non-padding) labels
    mask = tf.math.logical_not(tf.math.equal(labels, 0))
```

(continued)

Listing 10.15 (continued)

```
    # Use the mask to compute the loss only over non-padding tokens
    loss_seq = tf.keras.losses.sparse_categorical_crossentropy
(labels, predictions, from_logits=False)
    masked_loss_seq = tf.where(mask, loss_seq, 0)

    # Compute the mean loss
    loss = tf.reduce_sum(masked_loss_seq) / tf.reduce_sum(tf.
cast(mask, dtype=tf.float32))

    return loss
```

Table 10.1 The dimensions of a Transformer model

Component	Layer	Input dimension	Output dimension
Encoder	Input embedding	(batch_size, input_seq_len)	(batch_size, input_seq_len, embedding_dim)
Encoder	Positional encoding	(batch_size, input_seq_len, embedding_dim)	(batch_size, input_seq_len, embedding_dim)
Encoder	Multi-head attention	(batch_size, input_seq_len, embedding_dim)	(batch_size, input_seq_len, embedding_dim)
Encoder	Feed-forward network	(batch_size, input_seq_len, embedding_dim)	(batch_size, input_seq_len, embedding_dim)
Decoder	Input embedding	(batch_size, target_seq_len)	(batch_size, target_seq_len, embedding_dim)
Decoder	Positional encoding	(batch_size, target_seq_len, embedding_dim)	(batch_size, target_seq_len, embedding_dim)
Decoder	Masked multi-head attention	(batch_size, target_seq_len, embedding_dim)	(batch_size, target_seq_len, embedding_dim)
Decoder	Multi-head attention (with encoder output)	(batch_size, target_seq_len, embedding_dim)	(batch_size, target_seq_len, embedding_dim)
Decoder	Feed-forward network	(batch_size, target_seq_len, embedding_dim)	(batch_size, target_seq_len, embedding_dim)

We can then create the Transformer model as follows, leveraging encoder and decoder components:

Listing 10.16 Creating Transformer Model

```
import tensorflow as tf
from tensorflow.keras.layers import Input, Dense, Dropout,
Embedding
from tensorflow.keras import Model
from keras_nlp.layers import PositionEmbedding

def create_transformer_model(
    embedding_dim,
    num_heads,
    feedforward_dim,
    encoder_sequence_length,
    decoder_sequence_length,
    num_encoders=6,
    num_decoders=6,
 ):
    encoder_input = Input(shape=(None,), dtype="int64",
name="encoder_input")
    encoder_token_embeddings = Embedding(
        input_dim=vocab_size, output_dim=embedding_dim,
mask_zero=True
    )(encoder_input)
    encoder_position_embeddings = PositionEmbedding(
        sequence_length=encoder_sequence_length
    )(encoder_token_embeddings)
    encoder_embeddings = tf.keras.layers.Add()(
        [encoder_token_embeddings, encoder_position_embeddings]
    )

    x = encoder_embeddings
    for _ in range(num_encoders):
        x = TransformerEncoder(embedding_dim, num_heads,
feedforward_dim)(x)

    encoder_output = x

    decoder_input = Input(shape=(None,), dtype="int64",
name="decoder_input")
    decoder_token_embeddings = Embedding(
    input_dim=vocab_size, output_dim=embedding_dim,
```

(continued)

Listing 10.16 (continued)

```
mask_zero=True
    )(decoder_input)
    decoder_position_embeddings = PositionEmbedding(
    sequence_length=decoder_sequence_length
    )(decoder_token_embeddings)
    decoder_embeddings = tf.keras.layers.Add()(
        [decoder_token_embeddings, decoder_position_embeddings]
    )

    x = decoder_embeddings
    for _ in range(num_decoders):
        x = TransformerDecoder(embedding_dim, num_heads,
feedforward_dim)(
        x, encoder_output
    )

    decoder_output = Dropout(0.4)(x)

    final_output = Dense(vocab_size, activation="softmax")
(decoder_output)
    model = Model([encoder_input, decoder_input], final_output)
    model.compile(
        optimizer="adam",
        loss=masked_sparse_categorical_crossentropy,
        metrics=["accuracy"],
    )
    model.summary(line_length=120)

    return model
```

In the provided code, the model consists of two main parts: the encoder and the decoder. Both parts are composed of multiple layers, and each layer has its own specific role.

The encoder part consists of an input layer, an embedding layer, a position embedding layer, and multiple TransformerEncoder layers. The TransformerEncoder layers are where the self-attention mechanism is applied. The output from the encoder is used as inputs by the decoder. It also consists of an input layer, an embedding layer, a position embedding layer, and multiple TransformerDecoder layers. The TransformerDecoder layers also apply a self-attention mechanism, but in addition, they take the output from the encoder as input.

The model's final output is produced by a dense layer with a `softmax` activation function, resulting in a probability distribution across the potential output classes. The model is then compiled with the Adam optimizer (Kingma 2014), a loss function suitable for the task (in this case, `masked_sparse_categorical_crossentropy`), and accuracy as the metric. We will later demonstrate a case study on how to train this model and evaluate its performance.

It is important to note that modern large language models (LLMs) often employ a decoder-only architecture, which is autoregressive. For example, models such as GPT (Achiam et al. 2023) and Llama (Touvron et al. 2023) exclusively use this decoder-only architecture and are considered autoregressive models. This means they generate sequences one token at a time, from left to right. However, it is also not uncommon to encounter encoder–decoder architectures, such as the one used in the T5 model (Raffel et al. 2020).

10.8 Evaluation and Text Generation

10.8.1 Evaluation Metrics

Overview

Evaluating a language model can be accomplished using a combination of intrinsic, extrinsic, discriminative, and similarity metrics, complemented by interactive evaluation. Table 10.2 summarizes the evaluation metrics generally used to evaluate language models. Intrinsic metrics like perplexity measure a model's confidence in its predictions. Extrinsic metrics such as BLEU, ROUGE, and METEOR assess output quality in tasks like translation or summarization. Discriminative metrics such as AUC, precision, recall, F1-score, and accuracy evaluate how well a model distinguishes between correct and incorrect outputs. Similarity metrics, like embedding similarity, assess how close two text representations are in meaning. Interactive evaluation involves either using large language models (LLMs) or human judges to rate the quality of generated text based on factors such as fluency, coherence, and correctness. We will go into more detail on intrinsic and extrinsic metrics next.

Perplexity

The perplexity metric is a straightforward concept. It is essentially the exponential of the cross-entropy loss. Here is how it can be coded:

Table 10.2 Summary of evaluation metrics in language modeling

Category	Metrics	Description
Intrinsic metrics	Perplexity	Measures the uncertainty of a language model (Jelinek 1991). Lower perplexity means the model is more certain about its prediction
Extrinsic metrics	BLEU score	Measures the quality of text that has been machine-translated from one natural language to another (Papineni et al. 2002)
	ROUGE score	Measures the quality of summaries by comparing them to reference summaries (Lin 2004)
	METEOR score	Considers precision and recall, synonymy, stemming, and word order similarity to evaluate translation quality (Banerjee and Lavie 2005)
Discriminative metrics	Area under curve (AUC)	Measures the entire two-dimensional area underneath the entire ROC curve (from (0,0) to (1,1))
	Precision, recall, F1-score (classification)	Precision is the ratio of correctly predicted positive observations to the total predicted positives. Recall (sensitivity) is the ratio of correctly predicted positive observations to the all observations in actual class. F1 score combines precision and recall
	Accuracy	The ratio of number of correct predictions to the total number of input samples
Similarity metrics	Embedding similarity	Measures the cosine of the angle between two embedding vectors
Interactive evaluation	LLM-based eval	Asking a language model to evaluate the quality of the generated text
	Human eval	Involves having human evaluators rate the quality of outputs from a model based on various factors such as fluency, coherence, grammatical correctness, etc.

Listing 10.17 Calculating Perplexity

```
def calculate_perplexity(labels, predictions):

    # Cross entropy loss
    cross_entropy_loss = masked_sparse_categorical_crossentropy
(labels, predictions)

    # Calculate perplexity
    perplexity_score = tf.exp(cross_entropy_loss)
    return perplexity_score
```

BLEU and ROGUE Scores

BLEU and ROGUE scores are related to precision and recall. We will first need to understand how the precision and recall are calculated in a language modeling context.

Precision refers to how many n-grams in the candidate are also in the reference. This is defined as:

$$Precision = \frac{\text{Number of common } n - \text{grams}}{\text{Total number of } n - \text{grams in candidate}}$$

Recall refers to how many n-grams in the reference are also in the candidate. This is defined as:

$$Recall = \frac{\text{Number of common } n - \text{grams}}{\text{Total number of } n - \text{grams in reference}}$$

Let us now walk through an example of how precision and recall can be calculated for bigrams. Consider these two sentences:

reference: "*Spain are the champions in Euro 2024 and they are the favorites for World Cup*"
candidate: "*Spainish team are the champions champions in Euro 2024 becoming the favorites*".

Our first step involves constructing a count table for each bigram present in both the reference and candidate sentences. Afterward, we identify the minimum count from this table. Given the reference and candidate sentences above, the resulting count table would appear as follows:

Bigram	Count in reference	Count in candidate	Min count
("Spain", "are")	1	0	0
("are", "the")	2	1	1
("the", "champions")	1	1	1
("champions", "in")	1	1	1
("in", "euro")	1	1	1
("euro", "2024")	1	1	1
("2024", "and")	1	0	0
("and", "they")	1	0	0
("they", "are")	1	0	0
("the", "favorites")	1	1	1
("favorites", "for")	1	0	0
("for", "world")	1	0	0
("world", "cup")	1	0	0
("Spainish", "team")	0	1	0
("team", "are")	0	1	0

(continued)

Bigram	Count in reference	Count in candidate	Min count
("champions", "champions")	0	1	0
("2024", "becoming")	0	1	0
("becoming", "the")	0	1	0
SUM	14	11	6

Now, let us calculate precision and recall:

$$\text{Precision} = \text{Sum of Min Counts}/\text{Total Count in Candidate}$$

$$\text{Recall} = \text{Sum of Min Counts}/\text{Total Count in Reference}$$

From the table, we can see that the sum of the minimum counts is 6 (the total of the "Min Count" column). The total count in the candidate sentence is 11 (the sum of the "Count in Candidate" column), and the total count in the reference sentence is 14 (the sum of the "Count in Reference" column). So, the precision and recall are calculated as follows:

$$\text{Precision} = 6/11 = 0.55$$

$$\text{Recall} = 6/14 = 0.43$$

The following code implements the above calculations:

Listing 10.18 Calculating Precision and Recall Scores

```
from collections import Counter
import numpy as np

def count_ngrams(sentence, n):
    tokens = sentence.split()
    ngrams = [tuple(tokens[i:i+n]) for i in range(len(tokens)-n
+1)]
    return Counter(ngrams)

def calculate_precision(reference, candidate, n):
    ref_ngrams = count_ngrams(reference, n)
    cand_ngrams = count_ngrams(candidate, n)
    total = sum(cand_ngrams.values())
    correct = sum(min(c, ref_ngrams[g]) for g, c in cand_ngrams.
items())
    return correct / total
```

(continued)

Listing 10.18 (continued)

```
def calculate_recall(reference, candidate, n):
    ref_ngrams = count_ngrams(reference, n)
    cand_ngrams = count_ngrams(candidate, n)
    total = sum(ref_ngrams.values())
    correct = sum(min(c, ref_ngrams[g]) for g, c in cand_ngrams.
items())
    return correct / total
```

The calculation of the BLEU score involves several steps. First, for each n-gram size (up to a certain limit, typically 4), the precision is calculated as the number of correct n-grams in the machine translation divided by the total number of n-grams in the machine translation, as discussed above. These precisions are then geometrically averaged, and a brevity penalty is applied if the generated text is shorter than the reference.

The formula for the BLEU score is as follows:

$$BLEU = BP \cdot \exp\left(\sum_{n=1}^{N} \log p_n\right)$$

where:

- BP is the brevity penalty, which is 1 if the text length (c) is greater than or equal to the effective reference length (r), and $\exp(1 - r/c)$ otherwise
- N is the maximum n-gram size
- p_n is the precision for n-grams.

Here is the code that implements the BLEU score calculation:

Listing 10.19 Calculating Bleu Scores

```
def calculate_bleu(reference, candidate):
    len_reference = len(reference.split())
    len_candidate = len(candidate.split())
    brevity_penalty = np.exp(1-(len_reference/len_candidate))
if len_candidate < len_reference else 1
    precisions = [calculate_precision(reference, candidate, i)
for i in range(1, 5)]
    bleu_score = brevity_penalty * np.exp(sum(np.log(p) for p in
precisions) / 4)
    return bleu_score
```

The ROGUE-N score is essentially the same as the n-gram recall calculation we discussed earlier:

Listing 10.20 Calculating Rouge Scores
```
def calculate_rouge_n(reference, candidate, n):
    rouge_n_score = calculate_recall(reference, candidate, n)
    return rouge_n_score
```

10.8.2 Sampling Strategy

In language modeling, there are several sampling strategies available for generating the subsequent token in a sequence. The following table provides a summary of these strategies:

Sampling strategy	Description
Greedy decoding	Always selects the word with the highest probability as the next word in the sequence. It is deterministic and does not allow for much diversity in the output (Cho et al. 2014)
Temperature sampling	Modifies the probability distribution of the next word by applying a temperature factor. A high temperature (e.g., 1.0) makes the model output more random, while a low temperature (e.g., 0.1) makes the model more confident and conservative
Top-k sampling	At each time step, the model considers only the top k most probable words for the next word in the sequence (Fan et al. 2018). This introduces randomness in the output while still keeping it focused on high-probability words
Top-p (nucleus) sampling	Rather than choosing the top k words with the highest probability, it identifies the least number of words that have a combined probability that surpasses a specific threshold p (Holtzman et al. 2019). This allows for dynamic selection of the number of words considered at each timestep

10.9 Building an End-to-End Transformer Model

10.9.1 Question and Answering Bot

We will now look into a case study that involves training a Transformer model from scratch to create a question-answering bot. The scenario is as follows: given a natural language question and the SQL CREATE TABLE statement as context, the bot should be capable of generating the corresponding SQL query as an answer. Essentially, this is a text-to-SQL language model.

We will be utilizing the "sql-create-context" dataset provided by Hugging Face. This comprehensive dataset comprises 78,577 examples, each of which includes a natural language question, its corresponding context, and the associated answer. Here is an illustrative example:

```
{
    "question": "Please show the different statuses of cities and the
average population of cities with each status.",
    "context": "CREATE TABLE city (Status VARCHAR, Population
INTEGER)",
   "answer": "SELECT Status, AVG(Population) FROM city GROUP BY Status"
}
```

Let us first load the data from hugging face. The following code will achieve this:

Listing 10.21 Loading Text-to-SQL Data

```
from datasets import load_dataset

def load_sql_qna_data():
    dataset = load_dataset('b-mc2/sql-create-context')

    questions = [example['question'] for example in dataset
['train']]
    contexts = [example['context'] for example in dataset
['train']]
   answers = [example['answer'] for example in dataset['train']]

    data = list(zip(questions, contexts, answers))

    return data
```

Running the preceding code will load the data:

```
data = load_sql_qna_data()
```

As usual, we will split the data into train and test sets as follows:

Listing 10.22 Function to Split the Data into Train and Test Set

```
from sklearn.model_selection import train_test_split

def split_data(data):

    # Split the data into training and validation sets
    train_data, test_data = train_test_split(data,
test_size=0.2, random_state=42)

    return train_data, test_data
```

We will run this as:

```
train_data, test_data = split_data(data)
```

10.9.2 *Tokenization*

Next step in the pipeline is tokenization. This involves converting our text data into a format that our model can understand and process. We begin by preparing our text, adding specific tokens that signify the start and end of each sequence. We then train a byte pair encoder on all our text data, including contexts, questions, and answers from the training datasets. Special tokens are added to our tokenizer and padding is enabled to ensure all sequences are of the same length. Following this, we use the trained tokenizer to encode our texts, combining each question and its corresponding context into a single prompt. Lastly, we pad all our encoded sequences to a specified maximum length. This comprehensive tokenization process prepares our data for the subsequent stages:

Listing 10.23 Preparing Text and Tokenization

```
from tokenizers import Tokenizer, models

def prepare_text(text):
    text = '<START>' + text + '<END>'
    return text

def tokenization(train_data, test_data, sequence_length):
    # Unzip the data
    train_questions, train_contexts, train_answers = map(list,
```

(continued)

Listing 10.23 (continued)

```
zip(*train_data))
    test_questions, test_contexts, test_answers = map(list, zip
(*test_data))

    # Prepare text
    train_answers = list(map(prepare_text, train_answers))
    test_answers = list(map(prepare_text, test_answers))

    # Initialize a tokenizer
    tokenizer = Tokenizer(models.BPE())

    # Set padding token. Padding token must be zero.
    tokenizer.add_special_tokens(["<PAD>", "<START>", "<END>"])
    tokenizer.enable_padding(pad_id=0, pad_token="<PAD>")

    # Gather all texts
   all_texts = train_contexts + train_questions + train_answers +
test_contexts + test_questions + test_answers

    # Train the tokenizer
    tokenizer.train_from_iterator(all_texts)

    # Use the tokenizer to encode texts
    train_prompt_tokens, train_answer_tokens = [], []
    test_prompt_tokens, test_answer_tokens = [], []
    for question, context, answer in zip(train_questions,
train_contexts, train_answers):
        prompt = question + " context: " + context
        train_prompt_tokens.append(tokenizer.encode(prompt).
ids)
        train_answer_tokens.append(tokenizer.encode(answer).
ids)

    for question, context, answer in zip(test_questions,
test_contexts, test_answers):
        prompt = question + " context: " + context
        test_prompt_tokens.append(tokenizer.encode(prompt).ids)
        test_answer_tokens.append(tokenizer.encode(answer).ids)

    # Padding
    train_prompt_tokens = tf.keras.preprocessing.sequence.
```

(continued)

Listing 10.23 (continued)
```
pad_sequences(train_prompt_tokens, maxlen=sequence_length,
padding='post', value=0)
    train_answer_tokens = tf.keras.preprocessing.sequence.
pad_sequences(train_answer_tokens, maxlen=sequence_length,
padding='post', value=0)
    test_prompt_tokens = tf.keras.preprocessing.sequence.
pad_sequences(test_prompt_tokens, maxlen=sequence_length,
padding='post', value=0)
    test_answer_tokens = tf.keras.preprocessing.sequence.
pad_sequences(test_answer_tokens, maxlen=sequence_length,
padding='post', value=0)

    return train_prompt_tokens, train_answer_tokens,
test_prompt_tokens, test_answer_tokens, tokenizer
```

Let us now run the above function with a `sequence_length` of 50:

```
sequence_length = 50
train_prompt_tokens, train_answer_tokens, test_prompt_tokens,
test_answer_tokens, tokenizer = tokenization(train_data, test_data,
sequence_length)
```

10.9.3 Training

Now that we have converted raw texts into token IDs, we are ready to start training our Transformer model. The first step is to get three things ready: the 'encoder input', the 'decoder input' (shifted to the right), and the 'outputs'.

```
inputs = { 'encoder_input': train_prompt_tokens, 'decoder_input':
train_answer_tokens[:, :-1] }
outputs = train_answer_tokens[:, 1:]
```

Next, we will create a Transformer model configured with two attention heads:

```
encoder_sequence_length = 50
decoder_sequence_length = 49
vocab_size = tokenizer.get_vocab_size()
num_heads = 2
embedding_dim = 32
```

```
feedforward_dim = 64
model = create_transformer_model(embedding_dim, num_heads,
feedforward_dim, encoder_sequence_length, decoder_sequence_length)
```

Let us now train for 40 epochs:

```
from tensorflow.keras.callbacks import EarlyStopping, ModelCheckpoint

checkpoint_filepath = '../models/sql_qa_model.keras'
callback_es = EarlyStopping(monitor='val_loss', patience=5)
callback_mc = ModelCheckpoint(checkpoint_filepath,
monitor='val_loss', save_best_only=True, verbose=1)
history = model.fit(inputs, outputs, epochs=40, validation_split=0.2,
callbacks=[callback_es, callback_mc])
```

Here, to prevent overfitting and save computational resources, we employ two key strategies: Early Stopping and Model Checkpointing. Early Stopping monitors the validation loss and halts the training process if it does not improve after five epochs. Model Checkpointing, on the other hand, saves the model at the checkpoint file path whenever there is an improvement in the validation loss. This way, we ensure that we always have the best model saved, even if the training is stopped prematurely. The fit method initiates the training process, with 20% of the data reserved for validation, and the callbacks list includes our Early Stopping and Model Checkpointing strategies.

10.9.4 Prediction and Validation

Now that our model is trained, it is time to see how it does. We will use the test set we prepared earlier to check the model's performance. First, we need to load the model we saved:

```
model = tf.keras.models.load_model("../models/sql_qa_model.keras")
```

Next, we can leverage this model to generate predictions and subsequently validate these predictions using the test set. However, before we proceed, it is crucial that we first prepare the encoder and decoder inputs, as well as the corresponding labels:

```
encoder_input = tf.convert_to_tensor(test_prompt_tokens)
decoder_input = tf.convert_to_tensor(test_answer_tokens[:, :-1])
labels = tf.convert_to_tensor(test_answer_tokens[:, 1:])
```

We can then run the below code to produce the model predictions:

```
predictions = model.predict([encoder_input, decoder_input],
batch_size=8)
predictions = tf.convert_to_tensor(predictions)
```

To evaluate the perplexity score of our predictions on the test set, we can invoke our "`calculate_perplexity`" function as follows:

```
perplexity_score = calculate_perplexity(labels, predictions)
```

10.9.5 Text Generation

Ultimately, our objective is to generate text given a prompt using our trained model. It is important to note that our previous predictions utilized a decoder input that was shifted to the right. However, in a real-world scenario, we will not have access to the decoder input for the entire sequence. Instead, we will generate one token at a time, using the previously generated token as the new decoder input. Our end goal is to generate meaningful text (SQL query) given a prompt, which serves as the encoder input. The next code block illustrates this process:

Listing 10.24 Retrieving Answers Through Text Generation

```
def get_bot_answer(text, model, tokenizer, encoder_
sequence_length, decoder_sequence_length):

    text_encoded = tokenizer.encode(text)
    encoder_input = text_encoded.ids
    encoder_input = tf.keras.preprocessing.sequence.
pad_sequences([encoder_input],
maxlen=encoder_sequence_length, padding='post', value=0)
    encoder_input = tf.convert_to_tensor(encoder_input)
    encoder_input = tf.reshape(encoder_input, (1, 50))

    decoded_text = '<START>'
    for i in range(decoder_sequence_length):
        decoder_input = tokenizer.encode(decoded_text).ids
        decoder_input = tf.keras.preprocessing.sequence.
pad_sequences([decoder_input],
maxlen=decoder_sequence_length, padding='post', value=0)
```

(continued)

Listing 10.24 (continued)

```
        decoder_input = tf.convert_to_tensor(decoder_input)
        decoder_input = tf.reshape(decoder_input, (1, 49))

        prediction = model([encoder_input, decoder_input])

        # Greedy Decoding
        idx = np.argmax(prediction[0, i, :])
        token = tokenizer.decode([idx], skip_special_tokens=False)
        decoded_text += token

        if token == '<END>':
            break

    return decoded_text
```

The "`get_bot_answer`" function is essentially a chatbot response generator. The input text is first encoded and prepared for the model. The function then initiates a response generation process, where it iteratively predicts the next token of the response based on the current state of the input and the partially generated response. In the loop where the model makes predictions, the line `idx = np.argmax (prediction[0, i,:])` is selecting the word with the highest probability as the next word in the sequence. The function uses the method called "greedy decoding" for prediction, which simply picks the most likely next token at each step. The final output of the function is the generated response text.

Here is an example that highlights the power of Transformer models in action. Given a prompt that includes natural language and context, the model generates text that forms the structure of an SQL query:

```
-------------
PROMPT: What is the most recent year that China won a bronze? context:
CREATE TABLE table_name_77 (year INTEGER, bronze VARCHAR)
BOT: <START>SELECT MAX(year) FROM table_name_77 WHERE bronze = 1<END>
LABEL: SELECT MAX(year) FROM table_name_77 WHERE bronze = "china"
-------------
```

The prompt asks for the most recent year that China won a bronze, and the context provides the structure of the database table. The model's response is an SQL query that attempts to retrieve the desired information. Although the model's output does not perfectly match the label, it is remarkable that the model has begun to grasp the structure of SQL queries. This is particularly impressive considering that the Transformer model is trained from scratch and for only 40 epochs, without extensive hyperparameter tuning. It demonstrates the potential of Transformer models in understanding and generating structured query language.

10.10 Summary

- Transformer models eliminate sequential dependencies by processing all tokens in parallel, leveraging the self-attention mechanism.
- During training, transformers handle all input tokens simultaneously, but predict tokens one at a time during inference.
- Byte Pair Encoding (BPE) is a subword tokenization method that balances vocabulary size and sequence length by merging frequent character pairs.
- Key components of the transformer architecture include embeddings, encoder, decoder, and output layer.
- Embeddings consist of token embeddings for individual token representation and position embeddings for token position information.
- The encoder uses multi-head attention, feed-forward networks, and Add & Norm operations to process input embeddings.
- The decoder, with an additional sub-layer, applies a look-ahead mask and also processes encoder outputs to generate predictions.
- Evaluation metrics for language models include perplexity, BLEU, ROUGE, METEOR scores, and various similarity metrics.

References

Achiam, J., Adler, S., Agarwal, S., Ahmad, L., Akkaya, I., Aleman, F. L., Almeida, D., Altenschmidt, J., Altman, S., Anadkat, S., & others. (2023). Gpt-4 technical report. *arXiv Preprint arXiv:2303.08774*.

Ba, J. L., Kiros, J. R., & Hinton, G. E. (2016). Layer normalization. *arXiv Preprint arXiv:1607.06450*.

Banerjee, S., & Lavie, A. (2005). METEOR: An automatic metric for MT evaluation with improved correlation with human judgments. *Proceedings of the ACL Workshop on Intrinsic and Extrinsic Evaluation Measures for Machine Translation and/or Summarization*, 65–72.

Brown, T., Mann, B., Ryder, N., Subbiah, M., Kaplan, J. D., Dhariwal, P., Neelakantan, A., Shyam, P., Sastry, G., Askell, A., & others. (2020). Language models are few-shot learners. *Advances in Neural Information Processing Systems*, *33*, 1877–1901.

Cho, K., Van Merriënboer, B., Gulcehre, C., Bahdanau, D., Bougares, F., Schwenk, H., & Bengio, Y. (2014). Learning phrase representations using RNN encoder-decoder for statistical machine translation. *arXiv Preprint arXiv:1406.1078*.

Fan, A., Lewis, M., & Dauphin, Y. (2018). Hierarchical neural story generation. *arXiv Preprint arXiv:1805.04833*.

Gehring, J., Auli, M., Grangier, D., Yarats, D., & Dauphin, Y. N. (2017). Convolutional sequence to sequence learning. *International Conference on Machine Learning*, 1243–1252.

Hochreiter, S., & Schmidhuber, J. (1997). Long short-term memory. *Neural Computation*, *9*(8), 1735–1780.

Holtzman, A., Buys, J., Du, L., Forbes, M., & Choi, Y. (2019). The curious case of neural text degeneration. *arXiv Preprint arXiv:1904.09751.*

Jelinek, F. (1991). Principles of lexical language modeling for speech recognition. *Advances in Speech Signal Processing.*

Kingma, D. P. (2014). Adam: A method for stochastic optimization. *arXiv Preprint arXiv:1412.6980.*

Lin, C.-Y. (2004). Rouge: A package for automatic evaluation of summaries. *Text Summarization Branches Out*, 74–81.

Lipton, Z. C., Berkowitz, J., & Elkan, C. (2015). A critical review of recurrent neural networks for sequence learning. *arXiv Preprint arXiv:1506.00019.*

Papineni, K., Roukos, S., Ward, T., & Zhu, W.-J. (2002). Bleu: A method for automatic evaluation of machine translation. *Proceedings of the 40th Annual Meeting of the Association for Computational Linguistics*, 311–318.

Press, O., & Wolf, L. (2016). Using the output embedding to improve language models. *arXiv Preprint arXiv:1608.05859.*

Raffel, C., Shazeer, N., Roberts, A., Lee, K., Narang, S., Matena, M., Zhou, Y., Li, W., & Liu, P. J. (2020). Exploring the limits of transfer learning with a unified text-to-text transformer. *Journal of Machine Learning Research*, *21*(140), 1–67.

Sennrich, R., Haddow, B., & Birch, A. (2015). Neural machine translation of rare words with subword units. *arXiv Preprint arXiv:1508.07909.*

Touvron, H., Lavril, T., Izacard, G., Martinet, X., Lachaux, M.-A., Lacroix, T., Rozière, B., Goyal, N., Hambro, E., Azhar, F., & others. (2023). Llama: Open and efficient foundation language models. *arXiv Preprint arXiv:2302.13971.*

Vaswani, A., Shazeer, N., Parmar, N., Uszkoreit, J., Jones, L., Gomez, A. N., Kaiser, Ł., & Polosukhin, I. (2017). Attention is all you need. *Advances in Neural Information Processing Systems, 30.*

Williams, R. J., & Zipser, D. (1989). A learning algorithm for continually running fully recurrent neural networks. *Neural Computation*, *1*(2), 270–280.

The manufacturer's authorised representative in the EU is Springer Nature Customer Service Centre GmbH, Europaplatz 3, 69115 Heidelberg, Germany. If you have any concerns regarding our products, please contact ProductSafety@springernature.com

Printed and bound by CPI Group (UK) Ltd, Croydon, CR0 4YY
07/07/2026
02160907-0003